Terminal Care Support Teams
The Hospital–Hospice Interface

Edited by

R. J. DUNLOP
Medical Director, South Auckland Hospice, New Zealand

and

J. M. HOCKLEY
*Clinical Nurse Specialist (Terminal Care),
St Bartholomew's Hospital, London*

OXFORD NEW YORK TOKYO
OXFORD UNIVERSITY PRESS
1990

Oxford University Press, Walton Street, Oxford OX2 6DP
Oxford New York Toronto
Delhi Bombay Calcutta Madras Karachi
Petaling Jaya Singapore Hong Kong Tokyo
Nairobi Dar es Salaam Cape Town
Melbourne Auckland
and associated companies in
Berlin Ibadan

Oxford is a trade mark of Oxford University Press

Published in the United States
by Oxford University Press, New York

© R. J. Dunlop and J. M. Hockley, 1990

All rights reserved. No part of this publication may be reproduced,
stored in a retrieval system, or transmitted, in any form or by any means,
electronic, mechanical, photocopying, recording, or otherwise, without
the prior permission of Oxford University Press

This book is sold subject to the condition that it shall not, by way
of trade or otherwise, be lent, re-sold, hired out, or otherwise circulated
without the publisher's prior consent in any form of binding or cover
other than that in which it is published and without a similar condition
including this condition being imposed on the subsequent purchaser

British Library Cataloguing in Publication Data
Terminal care support teams: the hospital–hospice
interface
1. Terminally ill cancer patients. Care
I. Dunlop, R. J. (Robert J.)   II. Hockley, J. M. (Jo M.)
362.1'96994
ISBN 0-19-261915-2

Library of Congress Cataloging in Publication Data
(Data available)

Set by Footnote Graphics, Warminster, Wilts.
Printed by
Courier International, Tiptree, Essex.

Foreword
by
Dame Cicely Saunders

'A characteristic mixture of compassion and tough clinical science'—a comment in a review of a chapter on terminal care in a medical textbook aptly sums up two major characteristics of today's palliative medicine. The past two decades have seen much growth in both these areas and at the same time it has been shown how they can be translated and developed in a variety of settings.

Compassion is defined in the *Shorter Oxford English Dictionary* as pity that inclines one to spare or to succour, and it is the most appropriate word for care for those approaching the end of life. Much suffering can be spared by good communication, tailored to an individual patient's questions and needs, and of course by excellence in symptom control. Succour implies a move towards a positive outcome, the enabling of the potential that lies within the great majority of people who are battling their way through persistent disease. Anyone who has spent time in such work has seen strength and courage that neither patients nor family members realized were theirs. The unravelling of often long-standing problems, reconciliations, and new growth in relationships all frequently happen at a speed made possible by the pressure of crisis and limited time. This not only helps the patient to face letting go of life with acceptance and peace, but also gives important memories and new confidence to those who face the long and lonely journey of bereavement.

Tough clinical science is a major component of this type of succour, and the past two decades have seen a considerable volume of monitored clinical experience and research studies in both clinical and psychosocial areas. That palliative medicine is now an accepted speciality as well as a general challenge is the result of much hard work spread across the disciplines concerned. There is not, however, any excuse for complacency and the challenge is as much to those within the hospice movement constantly to question and improve their practice as it is to those who only meet these problems occasionally and have less experience of the present possibilities of relief.

Those first in this field were always concerned to reach patients wherever they were and to use the specialist centres as demonstrations of possibilities and principles for wider interpretation. Home care teams based in the early hospices were developed as complementary local services and were soon followed by teams with no back-up beds of their own. The peripatetic consulting hospital team arose at almost the same time as the new separate hospital wards or units and, with their experience shared with the existing ward teams, had a special potential for teaching. This was a particularly exciting way forward in the whole field of palliative medicine.

This book is an honest illustration of the difficulties as well as the possibilities of such an exercise. From hard-won and carefully observed experience it shows how a surprisingly small multidisciplinary team, developing from one pioneer and limited support, can bring both compassion and appropriate clinical skill to the most hard-pressed hospital wards and its many different staff members.

Anyone contemplating the introduction of such a team will learn the careful groundwork needed, the continual sharing and learning, and the support given and needed by a support team. The clear appraisal of the various challenges make fascinating reading and hopefully will make this excellent way of helping patients discover their own potentials at the end of their lives as widespread as it deserves to be.

March 1989

Preface

The expertise and example of the hospice movement has had a major impact on the lives of many terminally ill cancer patients and their families, but the majority of deaths still occur within other institutions, usually acute hospitals. The support care team, operating in an advisory capacity, has developed as an important alternative for the care of the dying. This book, which is based on the experience of several such teams, has been written to provide a framework on which others may base their efforts to plan and operate services for terminally ill patients in hospital.

We have reviewed the emotional, physical, social, and spiritual needs of terminally ill patients in hospital and their families, along with the variety of teams which have been established to meet these needs. Intergroup communication difficulties and the availability of high-technology facilities pose specific problems not often encountered in hospice experience and these issues are addressed. Information is provided about the palliative role of radiotherapy, oncology, and the pain clinic. The ethical basis for making decisions about treatment is discussed.

The role of the team in supporting and educating the carers, particularly the nursing staff, is also dealt with. The stresses that are encountered, both by the carers and the team, and strategies for minimizing these are examined.

To avoid repetition of earlier work, selected references on symptom control and home care are included in the Appendix.

London   R.J.D.
March 1989   J.M.H.

Contents

List of contributors		ix

1. The need for support teams
   1.1 Introduction — 1
   1.2 How the care of the dying became neglected — 2
   1.3 The modern hospice movement — 3
   1.4 Terminal illness and hospital — 4
   1.5 Why do patients die in hospital? — 5
   1.6 The needs of the patients — 6
   1.7 The needs of relatives — 8
   1.8 The needs of professional carers — 10
   1.9 Translating needs into team objectives — 11

2. Setting about meeting the need
   2.1 The palliative care unit — 13
   2.2 Terminal care hospital support teams — 14
   2.3 Some guidelines about planning a support team — 22

3. Selecting team members
   3.1 Introduction — 28
   3.2 Clinical nurse specialists — 29
   3.3 Doctors — 31
   3.4 Social workers — 34
   3.5 Chaplain — 36
   3.6 Secretary — 37
   3.7 Volunteers — 38
   3.8 Recruiting team members — 39

4. Team dynamics
   4.1 Introduction — 42
   4.2 Team goals — 42
   4.3 Patient referrals to the team — 44

	4.4	How support teams relate to patients and families	45
	4.5	The stress of working with patients and families	48
	4.6	The ethical basis for making decisions	50
	4.7	Communicating advice to the primary team	53
	4.8	What to do if advice is declined	55
	4.9	Other stresses on a support care team	57
	4.10	Relating together as a team	58
	4.11	Fulfilling the educational needs of team members	61
5.	Supporting the professional carers		
	5.1	Introduction	63
	5.2	Nursing staff	64
	5.3	Doctors	70
	5.4	Other professionals	73
6.	The pain clinic and palliative oncology		
	6.1	Introduction	75
	6.2	Pain clinic	75
	6.3	Palliative oncology	83
	6.4	Radiotherapy	83
	6.5	Hormone and chemotherapy	85
	6.6	Working with oncology services	90

Epilogue 93

References 94

Appendix 98

Index 99

Contributors

Wendy Burford RGN BTTA Certificate
Clinical Nurse Specialist (Terminal Care)
Brompton Hospital
London SW5

Robert J. Dunlop MB Ch.B. FRACP
Lecturer in the Care of the Dying
Support Care Team
St Bartholomew's Hospital
London EC1A 7BE

James M. G. Foster MB BS FFARCS
Consultant Anaesthetist, Pain Clinic
St Bartholomew's Hospital
London EC1A 7BE

Jacquiline Feld MA CQSW
Social Worker
Support Care Team
St Bartholomew's Hospital
London EC1A 7BE

Anne Goldman MA MB MRCP
Medical Director
Symptom Care Team
Department of Haematology and Oncology
Hospital for Sick Children
Great Ormond Street
London WC1

Tom Kerrane RGN BTA NAHC
Chief Nursing Officer to
National Heart and Chest Hospital
Bromptom Hospital
London SW5

Jo Hockley RGN SCM
Clinical Nurse Specialist (Terminal Care)
Support Care Team
St Bartholomew's Hospital
London EC1A 7BE

Rosemary Lennard MB Ch.B. Ph.D. MRCP
The Support Team
St Thomas' Hospital
Lambeth Palace Road
London EC1 7E

Judith Reddy RGN
Clinical Nurse Specialist (Terminal Care)
Support Care Team
St Bartholomew's Hospital
London EC1A 7BE

Dame Cicely Saunders DBE FRCP
Chairman and Founder
St Christopher's Hospice
Lawrie Park Road
London SE26 6D2

1

The need for support teams

1.1. INTRODUCTION

The main purpose of this book is to facilitate improvement in the care of terminally ill patients and their families in acute hospitals. The impetus for writing it came from the exciting changes we have observed since the formation of the St Bartholomew's Hospital Support Team. Dying patients no longer have to pluck up courage to interrupt a nurse, only to find that pain-killers are not available; they do not have to suffer their dark and terrible fears silently. Relatives have found support for their anguish, and relationships have been restored.

These changes have been made more exciting by the greater commitment of health professionals. Nurses have shown more confidence in their ability to care, and they now experience satisfaction from nursing dying patients. Doctors have become interested in the challenges presented by patients with incurable diseases. The sense of involvement has spread to other staff members, including ward clerks, technicians, and secretaries.

The therapies and insights developed within the hospice setting were the tools of change. But this book does not contain details about specific treatments, since these have been published already (see Appendix). Rather, this book is about how to adapt and use these tools within the hospital environment.

We have chosen to focus on hospital terminal care support teams because they seek to upgrade terminal care without assuming control of patients. This means that support teams can only advise the medical or surgical teams. Bringing the principles of hospice care alongside the disease-oriented hospital system is fraught with difficulties and frustrations. We have drawn upon the experience of several support teams to provide guidelines about coping with and overcoming these difficulties.

Terminal care support teams are known by a number of different names: support care team, palliative care team, symptom control team, and support team. For the sake of convenience, we will use 'support team' throughout the rest of the book.

Many support teams also provide a service for terminally ill patients in the community. Although some guidelines in this book are appropriate for home care, we have recommended specialist books on this subject in the Appendix.

Palliative care units can also effect changes in the management of hospital in-patients. They provide a hospice-like setting within the hospital, and frequently offer an advisory service. This book will provide some information about palliative care units.

By way of introduction, the next section of this chapter briefly examines the history of terminal care in hospital. The rest of the chapter reviews the needs of dying patients, their families, and professional carers. These needs are the reason why support teams were developed.

1.2. HOW THE CARE OF THE DYING BECAME NEGLECTED

The history of St Bartholomew's Hospital illustrates the change in attitudes toward the care of the dying. When Barts was founded in 1123 there was no difference in the meaning of 'hospital' and 'hospice'. The original Hospital house was a large hall in close proximity to the chapel. Rest and shelter were provided for the sick poor of London. The depth of commitment to the patients was evidenced by the Master of the Priory visiting them daily. As with many modern hospices, volunteers played an important role in the care of patients, and there was considerable reliance on charitable donations of money and food.

It is interesting to note that even at this time, the Hospital was associated with the possibility of cure. Early case studies document the resolution of a woman's swollen tongue when Rahere, the founder of Barts, placed a relic of the Cross upon it. The Book of the Foundation reports how 'some man joyed with jubilation at the remedy of his aching head, another for reparation of his going that he lacked'. In 1539 Thomas Vicary, a Surgeon, described the palliative function of Barts to King Henry VIII. The Hospital was 'for the ayde and comforte of the poore, sykke, blynde, aged, and impotent persones beying not hable to helpe themselffs nor havying any place certeyn whereyn they may be lodged, cherysshed or refresshed', to which he added 'tyll they be cured and holpen of theyre dyseases and syknesse' (Moore 1918).

The recognition of this potential for healing was not confined to the medical fraternity. Even in the 12th century, seriously ill people would undertake arduous journeys from all over England in the hope of a cure. The determination of some patients to pursue even the slimmest hope is still very evident.

It is frequently held that technological advances have prejudiced the care of the dying. However, attitudes of medical attendants had changed as early as 1544. The Royal Charter outlined the duties of the various staff involved in the care of patients at Barts. The role of the surgeons was 'to see if the patient were curable or not, so that none should be admitted who were incurable, none rejected who were curable' (Moore 1918). During the time of the Great Plague, physicians and surgeons distanced themselves by moving to the safety of the countryside. The care of patients was abandoned to 'the matron and apothecary' (Hector 1974).

1.3. THE MODERN HOSPICE MOVEMENT

Although Mme Jeanne Garnier opened the first hospice, especially for the dying, in France during the middle of the 19th century, it was not until the 1890's that Dr Howard Barrett, founder of St Luke's Home for the Dying Poor, tried to reach the public's interest in England (Saunders 1988) and steps were taken to redress the imbalance. The Irish Sisters of Charity established St Joseph's Hospice in close proximity to Barts Hospital, and the first in-patient was accepted in 1905. Since then, the Hospice has had continually to expand and it now has over 60 beds. Physical, psychological, and spiritual support became key elements of a theme of total care for the terminally ill patient and their family.

Hospice care took on a new dimension with the opening of St Christopher's Hospice in 1966. The restoration of the caring art of medicine was coupled with the sensitive application of the scientific method. In addition, there was a commitment to teaching the new-found knowledge and expertise. The number of hospices has increased considerably since then, and the contribution of the hospice movement to current medical practice cannot be doubted.

Now it would seem as though events have come full circle. The Barts support team represents a seed of the modern hospice movement in the oldest medieval hospital in England. Gradually the balance is being adjusted to incorporate the terminally ill once again.

1.4. TERMINAL ILLNESS AND HOSPITALS

The importance of the hospice movement has caused some people to loose their perspective on where people die. We have found that some hospital administrators and clinicians point to existing hospice services to avoid confronting the care of the dying in hospital.

Several studies have looked at where people die. In the retrospective study of Cartwright *et al.* (1973), 52 per cent of all deaths occurred in hospitals or institutions. The percentage of cancer patients dying in hospital has increased to 60 per cent since then (Lunt and Hillier 1981).

Patients with stroke, respiratory disease, and uncommon illnesses also tend to die in hospital. Up to 40 per cent of hospital deaths are related to ischaemic heart disease (Wilkes 1984). Cancer accounts for one third of hospital deaths. The prevalence of non-malignant diseases is an important difference from hospices.

Most patients who die at home will have spent some time in hospital. The majority of discharged patients are followed up by out-patient clinics, with at least half being seen on more than one occasion (Cartwright *et al.* 1973). Over 80 per cent of out-patients have symptoms that cause relatives considerable concern. This emphasizes that supportive care is necessary for out-patients as well as in-patients.

Many terminally ill patients have a relatively short stay in hospital. One third of patients will die within a week of final admission. However, 40 per cent will be in hospital for longer than a month (Wilkes 1984). This is considerably longer than the average length of stay for other patients in acute beds. Doctors often become concerned that dying patients 'block' the use of these beds and will want such patients discharged or transferred.

Terminally ill patients are less likely to die in teaching hospitals because the discharge rate is higher. Patients who die in teaching hospitals tend to be younger, 50 per cent being under the age of 65, compared with 30 per cent in district general hospitals (Cartwright et al. 1973). This is because younger patients are less likely to passively accept a terminal illness and will seek active treatment at a teaching hospital.

1.5. WHY DO PATIENTS DIE IN HOSPITAL?

Many people who want to care for the dying start with the premise that patients prefer to die at home. If this is not possible, 'home-like' surroundings would be the next best thing, and the hospital setting is certainly unsuitable. Mount et al. (1974) found that a number of health professionals, including nurses and social workers, felt that it was inappropriate for terminally ill patients to occupy acute hospital beds. The only group in favour were the terminally ill patients themselves.

One reason patients favour being in hospital is the sense of security. This reason is shared by relatives. At least 40 per cent of relatives feel that the hospital provides better care (Wilkes 1984). By dealing with fears, hospital support teams are able to allow more patients to die at home. However, the majority of patients still die in hospital (Bates 1985).

We undertook a prospective study of 100 patients who died in hospital (Dunlop et al. 1989). The reason why a patient died in hospital was determined by the support team member who had been most involved. We only found a minority of patients who wanted to leave hospital, and yet did not. Twenty-four patients wanted to die at home, or in some alternative to hospital. Although plans were set in motion to fulfil their desire, these patients deteriorated too quickly and were unable to travel.

Thirty-three patients died while receiving treatments such as radiotherapy and chemotherapy, or were being investigated for first presentation of advanced disease. Their deterioration was relatively unexpected. It might be thought that these patients were subjected to unnecessary interventions. However, several patients wanted to continue treatment when the doctors wanted to withdraw. These patients would ask us to convey their feelings to the doctors.

A significant number of patients (28) died in hospital because their relatives were unable to cope at home. There were three main reasons why this occurred.

1. The patient and relatives denied the illness and made no contingency plans before a crisis of symptoms precipitated admission. These patients and their families were often referred to the support team because their denial was considered 'inappropriate'.

2. Some patients were aware of the fact that the relatives were physically or emotionally unable to continue home care. These patients specifically requested admission to hospital. Often a close relationship between patient, family, and ward staff had built up over several admissions.

3. Several patients did not specifically choose to die in hospital but refused to consider transfers elsewhere. The ability to strike up new relationships and adapt to new environments steadily declines as an illness progresses.

Twelve patients were semi-conscious or comatose when referred and were too ill to be transferred elsewhere. Two patients died suddenly of cardiac arrest. One patient wanted to return home but had premonitory carotid artery bleeding. The doctors advised against discharge.

Having said that the majority of patients die in hospital, what are their needs there, and what needs can a support team therefore expect to meet?

1.6. THE NEEDS OF PATIENTS

Symptom control

Two prospective studies have looked at the symptoms of hospital patients with malignant and non-malignant conditions (Hockley *et al.* 1988; Hinton 1963). Both studies revealed that the majority of patients had distressing symptoms. Many patients had multiple symptoms. The relative frequency of symptoms such as breathlessness, anorexia, weakness, and pain are little different from hospice experience (see Table 1). However, sleeplessness was more common because of ward noise and routine turning.

Table 1. Symptom profile of 300 patients presenting to Bart's support team

Symptom	Incidence (%)
Pain	57
Anorexia	40
Constipation	27
Insomnia	21
Weakness/malaise	20
Dyspnoea	19
Nausea	18
Vomiting	14
Cough	13
Oedema	12
Sore mouth	10
Agitation/confusion	8
Dysphagia	6

Patients with non-malignant disease are just as likely to experience distressing symptoms. These patients, however, are less likely to have their symptoms relieved. Symptoms are poorly managed for several reasons: patients frequently accept symptoms as inevitable and do not complain; professionals fail to appreciate the symptoms, particularly the less obvious ones such as nausea, pain, and constipation; responses to symptoms are often inadequate, e.g. inattention to pressure areas and mouth care, and prescribing inadequate and infrequent doses of analgesics and laxatives.

Psychological and spiritual needs

Hinton (1963) found that at least 50 per cent of terminally ill patients in hospital described themselves as feeling depressed, particularly if patients were experiencing a number of distressing symptoms or were aware that they were dying. At least 75 per cent of patients dying in hospital will indicate to an observer that they are aware of impending death (Hinton 1963); 50 per cent will have talked to their relatives about dying (Hockley et al. 1988).

Anxiety was also common, particularly in patients who did not know their diagnosis. Mount et al. (1974) found that 80 per cent of patients thought they should always 'be told the nature of their disease'. At least two thirds of patients wanted to know the absolute truth about their prognosis. This was in complete contrast with the opinion and practice of the physicians. The doctors allowed their own fears about a potentially fatal illness to determine what they told patients.

Anxiety was also associated with some symptoms, particularly breathlessness. There was an association between anxiety and a tepid religious faith or dependent children. Other important causes of patient stress include the complex technology for treatment and diagnosis, and the prolonged periods of waiting for results of treatments or investigations (Degner and Beaton 1987). Patients in hospital are more likely to experience anxiety and irritability than patients in hospice (Hinton 1979).

At least 25 per cent of patients identify spiritual support as playing a major role in coping with their illness. However, up to 45 per cent of patients say that they are very unlikely to use a service providing spiritual guidance and support (Rainey et al. 1980).

Approximately 20 per cent of patients have no relatives or carers (Wilkes 1984). The other patients create and share the problems faced by relatives. The next section reviews relatives' needs, some of which are specific to the hospital setting.

1.7. THE NEEDS OF RELATIVES

Physical symptoms

Hockley et al. (1988) found that fatigue is the most common symptom experienced by relatives. In addition, up to 20 per cent of relatives developed a physical illness because of the stress of coping.

Psychological needs

Depression and anxiety result from the threatened loss of the patient. This stress is heightened when patients say they are probably going to die. Relatives find themselves changing the subject or trying to be optimistic, only to feel depressed because of the breakdown in a trusting relationship. Lack of information makes these emotions worse. Relatives receive conflicting statements about the patient's condition and prognosis (Hockley *et al.* 1988). Relatives may be desperate for more information but do not make this desire known. They are afraid to disrupt the routine of the doctors and nurses. Poor communication also results from the frequent changes in junior medical staff that occur.

We found that many relatives fear the patient will be discharged home (Hockley *et al.* 1988). This fear was related to the patient's distressing symptoms, and the perceived lack of general practitioner support. Relatives were also afraid of the patient falling or dying at home. At the same time, relatives felt guilty because they were not able to look after the patient at home. They saw this as letting the patient down.

Anger is often deflected away from the immediate situation on to social problems such as difficulties in transport and noisy neighbours. After the patient dies, up to 20 per cent of relatives will criticize the hospital as being uncaring (Wilkes 1984). It is less common for relatives to lodge a written complaint. Even so, complaints about the care of dying patients had constituted 8 per cent of the total number of complaints received by the Barts Hospital administration before the setting up of the support team.

Relatives' distress may be increased by the routine hospital procedures after a death. Requests for post-mortem may be accompanied by veiled threats of a Coroner's enquiry if the family do not agree. Many relatives find the task of picking up the patient's belongings in a Property Bag very upsetting. These problems all contribute to the impact of bereavement. Wright *et al.* (1988) have written an excellent review about the difficulties of those bereaved in hospital.

Social needs

Any progressive debilitating illness will have profound effects on the structure and integrity of the patient and family. Younger patients may loose earning power, with important financial consequences. Marital and family relationships will be disturbed by redistribution of roles, and by changes in the patient's sexuality. Social isolation, precipitated by the decline in strength and motivation of the patient, will compound these disruptive effects.

1.8. THE NEEDS OF PROFESSIONAL CARERS

Hospital staff are affected by patients' and relatives' problems. Being aware of staff needs is essential for a balanced approach to the care of the dying in hospital. We have briefly reviewed these needs in this section. Chapter 5 contains more detail about these needs and how to support the staff.

Nurses

Nursing staff are in close contact with terminally ill patients and their families. Forty per cent find the experience stressful while the remainder consider it 'rewarding but stressful at times' (Hockley 1989). Distressing physical symptoms, particularly pain and breathlessness, are very stressful. If nurses lack knowledge about or treatments for symptom control, they will try to avoid dying patients. Procedures such as dressing fungating wounds are not usually considered stressful unless there is insufficient time to perform them.

Many nurses find psychological care difficult. This applies to dealing with the anxious patient, the patient who asks about dying, and the needs of relatives.

Anxiety is the commonest emotional response to stress. First year nurses tend to be anxious about the difficulty of controlling their own emotions. More experienced students and trained staff are anxious about their responsibility toward the patient and relative. Depression and anger may occur, and nurses will sometimes feel 'overwhelmed'. It is rare, however, for these feelings to make nurses want to give up nursing or go off sick.

Nurses often lack support from other staff, including other nurses, supervisors, and doctors. When doctors find it difficult to make or communicate decisions about the management of patients, nurses bear the brunt of relatives' questions and anger. Disagreements about treatment decisions may also exacerbate stress, particularly when aggressive treatments are continued into the terminal phase and are seen as unwarranted (Degner and Beaton 1987). The converse may also apply, such as when patients with symptoms are ignored or when potential treatments are denied patients who are still relatively well.

The need for more teaching about the care of the dying is

frequently emphasized (Hockley 1989). First and second year nurses often wish to learn how to cope with their own feelings and reactions. They also want to learn how to talk to dying patients, not just answer difficult questions but facilitate ordinary conversation while performing a bed bath for example. Senior student nurses want to learn about how to deal with families, especially when a patient has died. Qualified nurses realize that they can influence treatment decisions and want more details about symptom control.

Doctors

Doctors find it very stressful making decisions about continuing on or stopping treatment of some patients who are likely to die (Degner and Beaton 1987). They experience anger and frustration whenever patients deteriorate despite treatment. Stress is increased by the feeling that other doctors are waiting to point out mistakes which might have contributed to the patient's decline. Doctors seldom discuss these feelings with each other.

Lack of knowledge about symptom control also contributes to a sense of helplessness. This results in terminally ill patients being ignored during ward rounds. It is little wonder that symptoms are under-detected, and that doctors over-report anxiety as a problem of patients (Wilkes 1984).

By default, junior doctors often have to assume the care of these patients, yet up to 50 per cent of junior doctors report that they have received inadequate training to enable them to cope with the care of the dying (Ahmedzai 1982). Junior staff find themselves having to act as go-between for ward nurses and consultants. The stress of having to report unwelcome problems to their seniors may be considerable.

1.9. TRANSLATING NEEDS INTO TEAM OBJECTIVES

No team can work without clear objectives. The needs found in the hospital provide the basis for the aims of any support team:

(1) to assist in the relief of distressing symptoms and to give emotional, social, and spiritual support to patients who have a terminal illness;

(2) to provide counselling and support to relatives and the bereaved;

(3) to provide support and advice to the staff caring for these patients; and

(4) to take part in education programmes on a multi-disciplinary basis.

The remainder of this book examines more fully how these aims can be realized.

2

Setting about meeting the need

The needs of terminally ill patients and families in hospital have been reviewed in the previous chapter. We will consider here the two principal developments of the hospice movement which have sought to meet these needs: the palliative care unit and the hospital support team. The concept of palliative care units is presented first, and is an overview based on features of several units. Readers who want more information about setting up and running a unit should refer to the excellent manual by Mount *et al.* (1974). In keeping with the theme of this book, the information given about support teams is more comprehensive. This chapter contains specific details about the setting up of four support teams and some general principles are drawn from these examples.

2.1. THE PALLIATIVE CARE UNIT

The palliative care unit is a purpose-built area within the hospital. Some units are in the same grounds but are physically separate from other hospital buildings. Alternatively, a separate ward or part of an existing ward may be devoted to the care of the dying.

Most units provide a homely atmosphere where careful attention has been paid to the decor of rooms. Carpeting, soft furnishings, restful colour schemes, and the use of plants and paintings all combine to produce a therapeutic environment. Background technology is kept to a minimum. Facilities are often available to allow relatives to stay overnight, make tea and coffee, and phone other members of the family. Separate catering facilities permit flexibility in preparing and serving meals.

A high proportion of trained nursing staff is characteristic of palliative care units. This allows more time for individualized physical and psychological care for patient and family. Medical input may be through a specific consultant attached whole-time, or working part-time in addition to a hospital or general practice appointment. In some cases, the primary team retains overall control and is responsible for visiting their patients. Many units have a chaplain and social worker attached part-time.

Continuity of care within the unit ensures ease of re-admission for out-patients. Many units offer respite care to enable families to enjoy a 1–2 week rest.

It is often not difficult to raise the capital to refurbish a ward or to establish a free-standing unit. However, it is much more difficult to meet running costs, particularly the salaries required to maintain a high patient/staff ratio. Inevitably, this restricts the number of beds, and units are only able to deal directly with the needs of a small minority of all patients.

A small number of beds means that an aggressive selection policy must be adopted. Otherwise, beds quickly become filled with patients who have slowly progressive disease and who cannot live at home. On the other hand, the transfer of patients who have only a matter of hours to live is upsetting to patients, family, and staff alike.

The presence of a palliative care unit may promote a 'dumping ground' philosophy in some doctors involved in acute medical care. When a dying patient presents to their service, or a patient already under their care deteriorates, they expect the patient to be moved to enable other acute patients to be admitted. If the patient is not transferred, there may be disappointment, even resentment —sentiments which detract from the patient's ongoing care in the ward. Palliative care units could therefore accentuate the idea that death must be hidden, kept apart, and not allowed to intrude on life.

However, the staff of such units are conscious of improvements in the management of terminally ill patients. They provide the opportunity for student nurses and junior medical staff to spend time on the unit, and the involvement of unit staff in teaching programmes. Some units extend their role to provide an advisory service for the other wards.

2.2. TERMINAL CARE HOSPITAL SUPPORT TEAMS

Where geographic, financial, or philosophical reasons mitigate against the establishment of a palliative care unit, the hospital support care team provides another alternative. The first such team was established in St Luke's Hospital in New York in 1975. The concept was carried over into the United Kingdom when, in 1976, Dr Thelma Bates, a Consultant Radiotherapist and Oncologist, conceived of establishing such a team at St Thomas's Hospital, London (Bates *et al.* 1981).

St Thomas's Hospital Terminal Care Support Team

Dr Bates met with several of the consultants and ward sisters throughout St Thomas's Hospital and confirmed her impression that there was an existing need. She then spent time obtaining the support of the District Nursing Officer, principal social worker, Dean of the Medical School, and Professor of General Practice. At the same time, sources of funding were established, particularly through the Hospital's special trustees and other charities. This enabled a detailed plan to be provided for the representatives of the medical and surgical services and the community health council. Having obtained approval in principle, a further six months was required before the team became operational. This time was spent setting up an office, developing job descriptions, and conducting interviews.

The team initially comprised a chaplain, social worker, nursing sister and two part-time voluntary doctors. Initial fears and anxieties on the part of doctors about loss of control over their patients, and potential psychological trauma to patients who were seen by the team rapidly dissolved. The ready acceptance of the team was attributed to the initial small number of members, their quiet and tactful manner, and the fact that Dr Bates was already a member of staff (Bates *et al.* 1981).

Despite being an advisory service, the team was able to facilitate a level of symptom control which enabled many patients to be discharged home earlier than was anticipated. The obvious advantage of such a service led to the appointment of two further nursing sisters to work in the community. Difficulties were experienced in relating to some general practitioners in the initial phase but a strict policy of respecting the wishes of these doctors overcame these.

Community care continues to be an important role of the team, with three nurses being employed specifically to maintain contact with out-patients. These nurses are involved in visiting patients at home but they also provide teaching for medical and nursing students, and district nurses. A 24 hour on-call service is provided by the team.

The team has maintained an average workload of approximately 300 patients per year. This constitutes an average of 20 in-patients and 40 out-patients at any one time. In-patients remain under the care of the primary medical or surgical team, and they are not transferred to specific beds. A full-time doctor and two in-patient nurses maintain a continued presence within the hospital, reviewing patients frequently and liaising with the primary teams. They are also involved in teaching medical students (Hoy et al. 1984), junior doctors, and nurses by way of formal and informal lectures and ward seminars. Research is another priority of the team; one of the doctors is employed as a Research Fellow.

The support team is also involved in operating a hospice day-care centre, the first of its kind in an acute general hospital in the U.K. This provides an opportunity for patients to enjoy the company of volunteers and other patients in a spacious relaxing environment.

St Bartholomew's Hospital Support Care Team

The Barts Hospital Support Team arose out of Jo Hockley's desire to adapt the model of the Team at St Thomas's Hospital. She had been a Ward Sister at St Christopher's Hospice for four years and felt she would like to bring her experience into the Hospital. With this vision came the courage to approach the nursing administration about the possibility of a Clinical Nurse Specialist post.

Miss Pam Hibbs, the Chief Nursing Officer, was very sympathetic toward the idea. However, it was not immediately possible to establish a new clinical post because of lack of funds. Money was available for research in terminal care and a project was completed which examined the needs of the dying patients, their families, and the nursing staff (Hockley 1988, 1989). The research clearly demonstrated that significant needs existed and were not being met. The results stimulated the setting up of a working party which comprised two Consultants, the Head Social Worker, Chaplain, the Chief Nursing Officer, and two Nursing Administrators. The possibility of setting up a multi-disciplinary Support Team was discussed, and agreement was reached in principle. The idea of establishing a separate ward for terminally ill patients was rejected because consultants might expect their patients to be transferred. It was felt this would reduce the incentive to improve terminal care overall.

Funds were made available for Miss Hockley to continue as a Clinical Nurse Specialist. Office space had been provided during the research project but it became necessary to find an office that was more central at Barts. Eventually it was arranged that the Nurse Specialist would share the Night Supervisor's office, which meant everything had to be cleared away and made tidy at the end of each day.

Dr Robert Davies, a Consultant Physician at Barts, recognized that the support team would not flourish without generalized acceptance from the physicians and surgeons. He therefore took on the role of promoting the team amongst his consultant colleagues.

One hundred and twenty patients were referred to the team in the first year. The clinical workload, coupled with teaching and the lack of specific secretarial support, proved demanding. The continued support of individual members of the nursing administration was especially important during this time. A social worker who had a particular interest in terminal care was another important source of personal support.

The working party decided not to appoint further nurse specialists until money was obtained for a team doctor. A doctor was needed for the team to gain credibility and to emphasize the medical significance of terminal care. Almost two years passed before a full-time doctor for the team was finally appointed. Napp, a pharmaceutical company specializing in drugs for terminal care, provided money for the first two years of the doctor's salary.

The team is now established in new offices. There is a full-time doctor, two specialist nurses, one part-time social worker, and a secretary at Barts, and two specialist nurses at the other major hospital in the district. These posts are now funded by the NHS. The number of referrals has continued to rise (325 in 1988) and every consultant in Barts has used the team. An after-hours service is not provided, although individual team members will leave a contact number when there is a patient with difficult problems on a ward.

The Hospital for Sick Children Symptom Care Team

Compared with adults, cancer in children is rare. The majority of children with cancer are treated at specialist centres, and the department at The Hospital for Sick Children, Great Ormond Street (GOS) is the largest. The paediatric oncologists at GOS have a strong tradition of caring for the physical and psychosocial needs of their patients from diagnosis, through treatment to either cure (up to 60 per cent) or relapse and ultimately death. For many years, this theme of continuity and total care had been provided by a team which included social workers, psychologists, play therapists, and teachers, as well as nursing and medical staff. Nevertheless, there were aspects of care which needed more attention: symptom care during treatment, terminal care, liaison, and education in paediatric palliative care.

Many children receiving therapy were experiencing nausea, vomiting, constipation, pain from procedures, needle phobia, and anxiety. Even though many patients would eventually be cured, these symptoms seriously disrupted their way of life. Nursing staff and particularly medical staff often focused on the intensive investigations and treatment of the cancer itself, and attention to these problems tended to take second place.

The extremely intensive, 'high tech' nature of paediatric oncology meant that the staff in GOS tended to be tuned to the needs of patients and families on active treatment. Changing gear to palliative care for a terminally ill child on the ward was often emotionally difficult. There were also practical problems finding the time and the continuity of staff needed for such patients.

The situation of having a terminally ill child with cancer at home is so uncommon that it may occur only once or twice in the lifetime of a family practitioner, and infrequently for a general paediatrician. Home care almost inevitably provokes considerable distress and feelings of inadequacy for everyone concerned. There seemed to be a need to provide a source of experience in dealing with the psychosocial support and symptom care for the family of the patient who had chosen to be at home, their primary health care team, and local hospital team.

Families frequently found it difficult to place the same confidence in the local hospital as they did in GOS. Misunderstandings or even small differences in approach between the various centres were worrying to families already under stress. It was often difficult to maintain good communication between GOS, the local hospital and the primary health care team, not least of all because of the wide geographical area from which the GOS patients come. It seemed likely that communication and the quality of care for the patients would be improved if staff from GOS could visit the local hospitals.

Although there are two children's hospices in the United Kingdom, their primary role has not been as a place for children to die but as a supportive facility for families with children who have progressive, often long-term, life-threatening disease. The study of symptom control and of training in palliative care in adults has not been paralleled in paediatrics. Although some principles of hospice care apply to children, it is not enough merely to extrapolate from experience in adults. The pharmacological handling of drugs and the types of malignancies affecting children differ from those of adults. Although children's symptoms overlap with those of adults, their relative frequency and importance will vary. The practical details of nursing care and the type of support needed by families with young children also differ from those of adults. It was recognized that a great need existed for paediatric hospice research and education.

Dr Ann Goldman conceived the idea for establishing a symptom care team in late 1984 whilst working in paediatric oncology and laboratory research. It was introduced to the oncology and haematology consultants and met with interest and enthusiasm. Basic outlines for the structure and roles of the team within the unit evolved rapidly.

The major hurdle was to find funding for staff salaries and expenses, including travel and running costs. The primary role of the team was to be clinical, and although some research was proposed, this was insufficient to attract grants from the large cancer research foundations. Eventually, over a year later, a fortunate link was established with The Rupert Foundation, a private charity dedicated specifically to the care of children with life-threatening disease.

A formal proposal for the team was drawn up with aims, *modus operandi* and estimated costing for an initial staff of one doctor, two nurses, and a part-time secretary. This was accepted, with funding for three years in the first instance, and the team began in early 1986.

The team has integrated well in the paediatric oncology department. Concerns about role definition and overlap, particularly with the social workers, resolved as it became evident that there was plenty for everyone to do.

Up to 5 newly diagnosed or relapsed in-patients are referred to the team each week and between 7 and 12 terminally ill children are being cared for at any one time at home. The team also maintains contact with the relapsed patients on treatment (currently about 35), those on treatment for the first time (currently about 150), and bereaved families that the team have cared for (currently about 80).

Brompton Hospital Support Care Team

In 1975, Tom Kerrane, Chief Nursing Officer of National Heart and Chest Hospitals undertook a study tour in the U.S.A. He met with a number of nurse specialists in various clinical areas, including terminal care. The calibre and performance of these senior nurses confirmed his view that the nurse specialist had an important contribution to make as part of the health care team in the general hospital or specialist centre.

Following his return, Mr Kerrane prepared an outline proposal for discussion within the Brompton Hospital, a 320-bed postgraduate cardiothoracic centre. Meetings with immediate senior nursing colleagues identified the terminally ill patients as a priority area for the possible appointment of a nurse specialist. This group of patients had grown in number but their care was not up to the standard expected from a major teaching centre. Two reasons were identified for the deficiencies in care. The main focus of nursing care was on the acute cardiothoracic patients. Their needs tended to overshadow those of the patients needing terminal care. In addition, there was a perceived lack of nursing expertise available to provide a comprehensive service for patients with limited life expectancy.

Contact was made with Dame Cicley Saunders (Medical Director of St Christopher's Hospice). She gave considerable encouragement and support to the plans to introduce a nurse specialist for terminal care.

Initial discussions with senior medical staff at Brompton Hospital were not fruitful. Although there was some recognition of the need for key nurses to provide extended care in specific areas, it was felt that the existing ward sister grade should meet these needs. Fortunately, a new young consultant physician was appointed at this time who had interests in lung cancer, symptom control, and terminal care. He responded quickly and positively to the proposals, and his support obviously influenced his medical colleagues.

There was some opposition to the idea from nursing staff. Some ward sisters were wary and suspected that a nurse specialist would interfere with and detract from their roles.

The Medical Social Work and Occupational Therapy Departments were also consulted before a job specification for a nurse specialist was finally agreed. The chief administrator for the Hospital gave the project his support, and funding was obtained from the Elizabeth Clarke Charitable Trust.

In April 1978, Wendy Burford was appointed Clinical Nurse Specialist to co-ordinate a multidisciplinary team with the aim of meeting the total needs of terminally ill lung cancer patients. Her role also included attending the oncology clinic and being involved with patients during the active treatment phase, particularly patients with small-cell lung cancer. Teaching commitments with the Department of Nursing Studies and the ward areas were also established.

There was no doctor specializing in terminal care attached to the team. The patients remained under the care of the consultant to whom they were originally referred. All physicians agreed that the nurse specialist could see their patients and follow them up in the local community with the permission of the general practitioners and district nurses.

In general, the nursing staff were supportive, although some of the staff were suspicious because of the innovative nature of the post. An easy working relationship was quickly established with the medical social workers when they discovered that the nurse specialist would not take over their role but complement it with nursing knowledge. Weekly meetings with the hospital chaplains were used to discuss the patients with whom the nurse specialist was working. Miss Burford found that her prior knowledge of the hospital and staff was a valuable advantage. However, this did not prevent a feeling of isolation because she was the only clinical nurse specialist in the hospital and did not have a peer group to relate to.

There has been a continual increase in the workload. This led to the appointment of a second nurse in 1981, which enabled the brief of the team to be extended to include patients with life-threatening non-malignant conditions.

From the time of her appointment, Miss Burford felt it would be important to develop an area within the medical unit where the needs of terminally ill patients could be more easily met, away from the activity and stress of a busy medical ward.

After much discussion it was agreed that an eight-bed area could be converted to provide a six-bed unit; a legacy of £7000 helped to cover the cost of conversion. The unit opened on 27 July 1980. Staffing was provided from the team of the medical ward where the unit was situated. It was noted that pain and symptom control was monitored and dealt with more promptly in the Continuing Care Unit.

The Continuing Care Unit closed in October 1987 due to financial constraints imposed on the National Health Service, but was able to re-open in July 1988. The service to patients in other medical wards has continued throughout this time.

2.3. SOME GUIDELINES ABOUT PLANNING A SUPPORT TEAM

It will be clear from the above descriptions that there is no single blueprint for establishing a hospital support care team. However, there are some general principles which will help guide those who may be planning to set up such a team.

Someone with vision

Most teams have started from the enthusiasm and insight of individual doctors or nurses who have been moved by the special needs of the terminally ill. These individuals have often had contact with the hospice movement. Other health professionals may interpret the proposals for a terminal care team as an attempt at 'empire building'. It is vital therefore that at least one person has the necessary vision and stamina to survive the inevitable delays, frustrations, and set-backs that characterize the planning phase; it may take several years from the inception of a project to a team becoming operational.

Occasionally, support care teams have been established as the result of regional administrative directives. Some hospitals have been instructed to improve the care of the terminally ill. Directives can readily be misconstrued as criticism of existing services. A visionary will be just as necessary to ensure the project survives the suspicion and distrust.

Mobilizing support

There may be little difficulty in finding other people willing to lend moral and practical support to setting up a team. If the visionaries are not part of the power structure of the hospital then it will be essential to lobby doctors, members of the nursing management, and administrators.

Mention has been made of the negative feelings to proposals about establishing a team. The fact that money is being used to improve terminal care at a time when funding of established services is being decreased will add bitterness to the sense of threat and criticism. Opponents may actively undermine the team by spreading misinformation and witholding patient and family referrals. Several teams have felt that more time should have been devoted to addressing such sentiments during the planning phase.

The question may arise about whether the needs of the dying can be met in hospital. It has been argued that acute wards are not suitable for terminally ill patients because the necessary standard of care cannot be provided (NWTRHA 1987). This argument could seriously weaken support for plans to set up a team. There is some data to confirm anecdotal evidence that change can be effected. Studies are currently under way which should provide more definitive answers.

The National Hospice Study compared hospital-based hospice units, hospice home care, and selected hospital wards in the United States (Greer *et al.* 1986). Hospital-based hospices provided better control of pain and other symptoms, and relatives were more satisfied with the care. The differences between settings were relatively small, however, and there was no significant difference in relatives' perception of patient quality of life.

The results did not convey the fact that the quality of conventional hospital care improved during the study. The change occurred because hospital care-givers were under review and their motivation to improve standards increased (D. S. Greer, personal communication). This suggests that heightening awareness of deficiencies in care can stimulate change—the presence of a support team can achieve this.

Parkes' longitudinal study (1985) suggested that education can improve pain control in hospital. In the initial study, the small number of hospice patients experiencing severe pain prior to death contrasted sharply with the hospital experience. During a ten-year period, pain control in the hospitals near St Christopher's Hospice improved to the extent that there was no longer any significant difference. Parkes felt that the improvement represented an effect of the Hospice's education programme. Many doctors and nurses from the study hospitals had attended teaching sessions.

Although patient care improved, Parkes found that the care of relatives was less satisfactory in hospital. He suggested that attention to relatives' needs was difficult to translate into the hospital setting. Indirect evidence indicates hospital support teams can improve relatives' care. We reviewed letters of complaint received by the Barts administration. Before the introduction of the Barts Support Team, 8 per cent of the total number of complaints came from relatives of dying patients. These complaints declined steadily to less than 1 per cent after four years of the team being in operation (Hockley *et al.* 1988).

Identifying the needs

Although the weight of evidence about unmet needs of patients, families, and staff needs is considerable, sadly it may still be necessary to convince some administrators and other influential members of staff that the terminally ill do not receive adequate care. Several people have undertaken projects to assess their local situation. This need not involve a complex, structured research programme; there are a number of simple tools available. Quite apart from the collection of data, these projects can serve to improve the acceptance of the support team in the early operational phase. Hospital staff will become familiar with the investigator before he or she begins operating in an advisory capacity as a member of the support team.

Identifying services already involved in meeting needs

Some effort should be made to meet with general practitioners, district nurses, hospice staff, home care teams, bereavement groups, and other services which are providing care for the terminally ill within the district. This is particularly important if a support team is to be funded by the NHS; many other services are funded charitably and may feel their existence is threatened by the new service. Concerns may also be eased by allowing existing services to be represented on a planning committee.

Even with good preparation, difficulties will arise because of perceived and actual similarities in roles. Doctors may find it difficult to accept that nurses will provide advice about medications. Social workers are likely to be unsettled by the appointment of nurse specialists who have a specific counselling role. Doctors and nurse specialists in radiotherapy and oncology units may feel threatened if the support team develops separately.

Finding funds

Obtaining money for a support team can be very difficult. Insufficient funds frequently limit the size of a team. You should expect negotiations to take a long time, and to be punctuated by frequent setbacks.

Ideally, the salaries of team members should be provided by the local health authority but this is rarely realized in the first instance. Occasionally, the regional health administration will make funding available to implement directives. It may be possible to obtain short-term funds from this source for researching needs.

More commonly, short-term funding has to be obtained from charitable trusts such as Macmillan Cancer Relief Fund. Some of the other trusts which have provided money were mentioned in the previous description of specific teams. Drug companies have sometimes provided priming money. Ultimately, health authorities need to take over after two or three years.

The availability of money from non-hospital sources may make it difficult to gauge the true level of administrative support. The administration may pay lip-service to the principles of terminal care. But the true depth and sincerity of support will only become apparent when the short-term funding expires. A lack of concrete commitment will have a devastating effect; the withdrawal of financial support was instrumental in the collapse of the Charing Cross Hospital Team (Herxheimer *et al.* 1985).

We have found that donations and bequests provide some money to cover administrative and minor capital costs, but are extremely unlikely to provide for salaries.

Planning for the practical needs of a support team

Careful thought should be given to the name of the team as this can affect how the team is perceived. One team was criticized for being called 'special care team'—other staff wondered why the team was so 'special'. Two clinical nurse specialists were identified as 'care of the dying sister'. Only patients who were actually dying were referred and the nurses felt their skills were underutilized.

We have found the title 'support care team' to be useful. The initials can be interchanged with 'symptom control team'. This permits flexibility when being introduced to a patient. The former name immediately conveys a sense of comfort to patients with psychosocial problems, whereas the latter name reassures patients with symptoms who may be hesitant about the psychological connotations of 'support'.

The provision of adequate office space is important. The physical environment of the office can have important effects on team morale and stress. Space is very often at a premium in existing hospitals, and office space is likely to be small and cramped. An increase in the number of team members will make the effect of office size more pronounced.

An office which is geographically distant from hospital wards will create a sense of isolation, particularly if there are no other full-time team members. Maintaining a profile on the wards will be difficult. Consequently, hospital staff will be less familiar with the team, and fewer referrals will be made. The prospect of a significant journey will deter retreat to the relative peace of the office after a stressful experience. An office by a busy corridor will increase contact with other hospital staff but may not allow privacy for team meetings.

It is important to ensure an adequate number of phones. In

general, there should be one phone line per team member. A direct line which bypasses the hospital switchboard is helpful for relatives and out-patients. When a number of patients are referred from outside the district, the capability to make long-distance calls will facilitate liaison with other teams and hospices. An answerphone will improve communication, particularly if there is no secretary to take calls during the day; some relatives appreciate being able to leave a message after hours when there is no 24-hour service.

There are two other steps which are essential to the planning process; developing job descriptions, and selecting team members. Because of the importance of these related steps, we have devoted the next chapter to them.

3

Selecting team members

3.1. INTRODUCTION

The examples of support teams presented in the previous chapter give some indication of the variety of responses that have appeared to fit the needs of dying patients and their families in different hospitals. In almost all support teams, clinical nurse specialists will form the backbone. One or more doctors, a social worker, secretary, and the clergy may also be represented in many different combinations. Volunteers, social work assistants, bereavement counsellors, and health visitors also work as an integral part of some teams.

It is not uncommon for teams to start with one member, indeed many small hospitals have only a single nurse specialist (Morris 1981). For the purposes of compiling data about services for the terminally ill, solo nurse practitioners are not counted as 'teams'. However, it is vitally important to recognize that terminal care nurse specialists cannot effectively operate in isolation, and need a network of colleagues—nursing, medical, and other—to provide additional advice and support.

Each putative member of a new team should have a clearly defined role. Clear-cut practical job descriptions are essential to the process of selecting team members; applicants require guidelines on which to base their decision to apply for the post, and members of the interview panel must have an adequate understanding of the job in order to assess the suitability of applicants. Once staff are in post, a realistic understanding of each person's role will make for greater team effectiveness and harmony.

The following descriptions of the tasks, responsibilities, and skills required of team members are based on the collective experiences of various teams visited and interviewed by the authors. These profiles are not intended to be prescriptive but can be used as a basis for planning job descriptions. Where possible, these descriptions should be supplemented by a visit to teams working in similar circumstances before deciding on the best formula for local needs.

3.2. CLINICAL NURSE SPECIALISTS

The majority of nurses on support care teams are employed full-time; nurses working part-time usually complement full-time colleagues. A nurse specialist will often be the main co-ordinator by virtue of being the only full-time team member. In larger teams, the majority of team members will be nurse specialists, which parallels the fact that nurses constitute the greater proportion of care-givers within the hospital.

The position of clinical nurse specialist implies a high degree of knowledge and expertise. The credentials of any person employed must evoke respect from other nursing and medical staff. The nurses who have been appointed manifest considerable variation in previous training and experience. There is a general tendency for them to have a further qualification other than RGN. Most nurses who have had prior experience as ward sisters or district nursing sisters find this to be an advantage. District nurse or health visitor qualifications may be an absolute requirement if members of the team are to work in the community.

Experience in caring for the terminally ill has not necessarily been a selection criteria. Less than 50 per cent of nurses working in support teams have had more than twelve months experience working in a hospice prior to their appointment. This may reflect the small number of nurses with hospice experience who apply for such posts—many nurses take up jobs in hospices because they are dissatisfied with the quality of care provided in hospitals and may not want to return. Clinical nurse specialists have been able to work effectively in this field with a minimum of prior hospice experience, but a dedication to learning the principles is essential.

Most support team nurses have gained their experience after their appointment, either by attending an ENB course in 'Care of the Dying', or by obtaining limited work experience in one or more hospices. The value of this practice has not been established. Those nurses who had prior hospice experience felt this had been essential for enabling them to undertake the advisory and teaching roles. Two thirds of support teams felt that more hospice experience would have been beneficial in reducing the stress associated with advising hospital doctors and general practitioners about symptom control. Others felt that although working in hospice provided considerable experience in symptom management, it did not prepare them for the major stresses of trying to apply these principles within the hospital setting (Hockley, unpublished data).

Health visitors can be employed on support teams. Their community-based experience will provide them with the necessary credentials for working alongside district nurses in home care. They have also conducted bereavement care and follow-ups.

Many nurses wear a sister's or nursing officer's uniform in the hospital or community. This has the advantage of immediately clarifying the status of the nurse to the patient, family, and other staff. Wearing everyday clothes and a name badge identifying status has the advantage of emphasizing the advocate role of the specialist on behalf of the patient and family. Either way, maintaining a distinction between the ward nursing staff and the nurse specialist allows one to teach and advise without taking over nursing procedures, although there are times when providing practical help is necessary.

Job descriptions normally require pathways of responsibility and accountability to be defined. Given the hierarchical structure of acute hospital services, these pathways will frequently involve senior nurses and other administrators who are physically remote from the team. They may be a valuable support to individual members of a team, particularly during the formative stages. Considerable stress may result when support is not forthcoming at this level. Supervisors should be committed to the principles of terminal care, the role of the nurse specialist, and the multi-disciplinary approach, in addition to their administrative skills. If they are not able to represent the team at management level, it will be necessary to recruit individuals who can.

Community nurse specialists have often been made responsible to a different nurse manager from the other nurses on the support team. This diffusion of accountability may produce significant tensions, as happens when the community nurse specialist is expected to perform 'hands on' care to supplement a shortage of district nurses.

Whether on the wards, out in the community or in the school of nursing, teaching plays an important part in the support team nurse's role. Having the skill and enthusiasm to pass on knowledge can be an asset to the daily practice of the nurse specialist. Quite apart from educating other carers, teaching can have a recharging effect in what may otherwise be, at times, an emotionally draining job.

Currently, there are few long-term career prospects if a nurse chooses to specialize in this area of terminal care. There may be considerable pressure for the senior nurse on the team to provide stability by remaining in post for a number of years. After 2–3 years as a nurse specialist on a support team, some people need the challenge of a different job. This means moving to a different field of nursing, or continuing in terminal care within a hospice or the community. Many hospices are looking for tutors with experience in terminal care to help run their courses. The community option is difficult if one does not have a district nurse or health visitor qualification. The courses require a two year commitment to a local health authority, with a return to general rather than terminal care nursing. Obtaining charity funding has allowed some people to reduce the health visitor course to one year.

3.3. DOCTORS

The majority of support teams have at least one doctor. Teams without a doctor find that they lack a degree of credibility within the hospital, particularly with the medical staff. Non-medical team members will be stressed by having to carry the responsibility of advising on medical issues.

It is difficult to establish full-time posts for doctors. This reflects the higher salaries of doctors and the relative dearth of doctors with terminal care experience—the rapid increase in the number of hospices has accounted for most of the doctors with hospice training. Maguire (1985) suggested that there may be some advantage in working part-time. More time is available to pursue other clinical interests, which reduces the stress inherent in terminal care. This consideration is important given that the team doctor's role is to advise on rather than take over care. Specialist training in other fields, including hospice, will have involved the doctor exercising responsibility for patients. It can be extremely difficult for a doctor to relinquish this. Negotiating every investigation or change in medication can be disheartening and humiliating, especially if the primary team declines to follow suggestions.

There is a wide variety of specialist backgrounds represented by the doctors working with support teams: oncology, radiotherapy, psychiatry, clinical pharmacology, anaesthetics, general medicine, surgery, and general practice. The common denominator is a deeply felt concern for improving the quality of life for terminally ill patients and families.

Whatever the clinical background, it will be important to attend specialist courses in terminal care or spend time attached to a hospice. The credibility of the doctor, particularly if he or she does not hold a specialist qualification, will depend on being able to control difficult symptoms, especially when the doctors of the primary team have strong emotional ties with a terminally ill patient. If the patient is a personal friend or a prominent citizen, the primary team will closely monitor what is done. If one is able to help, the 'specialist' barriers will be broken and the team will be more widely accepted and used.

Several teams have a doctor available for advice on a voluntary basis; the doctor usually attends the main multi-disciplinary meeting each week. This arrangement may work well in a small hospital, particularly when financial constraints limit the size of the team, but will prove unsatisfactory if the doctor usually has a heavy case load or is not able to attend team meetings. The resultant stress on the other team members can be greater than the stress of having no doctor at all. The difficulty of this arrangement was exemplified by the Charing Cross Hospital Terminal Care Support Team (Herxheimer *et al.* 1985).

One of the important responsibilities of the doctor is to provide expertise on difficult symptom control problems to other members of the team. Nurses can gain the experience necessary to advise on medication changes for many problems. But this experience may not always take into account subtle changes that may be necessary to avoid patients developing side-effects as a result of other medical conditions. The doctor may provide insight into the rarer manifestations of cancer and cancer treatment. The medical management of non-malignant conditions, such as manipulating diuretics and vasodilators in heart-failure patients, may require the specific expertise of a doctor. On occasion, the doctor will need to review investigations for difficult diagnostic problems or review response to radiation and chemotherapy, particularly if chemotherapy is being administered by non-oncologists.

Another important responsibility of the doctor is to debate medical management decisions with some primary teams. Medical teams will often respond readily to the advise of the nurse specialist. Problems arise when diagnostic and therapeutic possibilities are not clear cut or include high risk/low benefit options. If the doctor avoids this responsibility, greater stress is imposed on the other team members.

Some patients find it difficult to accept the advice and attention of the nurse specialist on the team. It is not uncommon for the nurse specialist to spend considerable time in resolving fears about the use of morphine, only to have the hard work shattered by a careless comment about addiction from one of the members of the primary team. It can be very helpful if the support team doctor lends his or her authority to the situation. But this should be done in such a way as to emphasize rather than override the status of the nurse specialist.

The non-medical members of the team will find it helpful if the doctor can provide some understanding about how doctors make decisions. Non-medical staff are not always aware of the unique stresses faced by the medical profession and this may lead to misinterpretation of decisions about patient management. The effectiveness of the team will be enhanced if anger is redirected into sympathetic understanding (not to be confused with agreement).

The teaching role of the doctor will depend on how many sessions are available. The doctor will often play a major role in teaching medical students and junior doctors; at consultant level, forums such as medical and surgical grand rounds may be used. Where possible, other members of the team should be involved to exemplify the multi-disciplinary approach.

The impetus for clinical research and publication may need to come from the doctor on the team. Nurses are becoming more aware of their ability to contribute in this regard because of the increasing number of research nurse posts. Even so, nurses' training has in the past suppressed initiative and this may need to be overcome.

The doctor, and the other members of the team, should be alert to the way in which the interest of the doctor may impose on and limit the functioning of the team. An anaesthetist might emphasize the role of nerve blocks for pain to the extent that this could lead to the team functioning predominantly as a pain clinic. Involvement with patients who have other symptoms or psychosocial needs could decrease. A strong interest in acute and chronic benign pain may also be unbalancing. A doctor who is also attached to oncology or radiotherapy may receive a number of referrals from these specialities. However, referrals from physicians, surgeons, and general practitioners may be reduced because of a fear that their patients may be 'taken over'.

Recruiting and retaining doctors for hospital support teams can be difficult. The relative lack of prospects for career advancement will be a consideration for some doctors. This applies particularly to senior registrar, clinical assistant, or lecturer posts because of the few options beyond these levels. Also, there is a tendency for many hospital doctors to perceive palliative care as an inferior branch of medicine. This perceived lack of credibility and fear of limited job prospects will hopefully change as more consultant and professorial posts become available. The fact that some colleges are prepared to recognize time spent training in palliative care will also be constructive. Meanwhile, team stability will be threatened by the potentially frequent turn-around of medical input to the team.

3.4. SOCIAL WORKERS

Social work has an important complementary role in terminal care. The medical social worker's task has traditionally been to help families with social, emotional, or practical problems. Smaller teams may have to rely on working with the individual social worker allocated to the primary team caring for the patient. An alternative is to liaise with a hospital social worker who has a special interest in the care of the terminally ill. A number of support care teams have a social worker specifically attached. Even then, we have found it helpful for the primary team social worker who is already seeing a patient, and who wishes to stay involved, to remain the key social worker.

The social worker can assess how physical, emotional, and social factors link together to influence the patient and family. Using counselling skills to assess a patient and the holistic approach to caring are certainly not the perogative of the social worker. But the social worker can contribute by recognizing that every patient is part of a social system which influences how he deals with his illness. Social system refers to the patient's cultural, religious, and social grouping, and includes other factors such as age and sex. It also acknowledges the patient's role in the family and his peer group. The social worker's contribution can increase the team's understanding of a patient and his pain.

Symptoms can be a message—part of the family's context of pain. The patient is not just an individual with problems and symptoms but a member of a family whose actions interact with his. Recognizing the patient as part of his family can be extremely helpful to the team since it provides the possibility of using that family's strength and resources, and reduces the team's need to be the 'sole saviours'.

The diagnosis of terminal illness can lead to personal confusion and isolation within a family. Family members may have to take on unfamiliar roles, such as a wife taking on business responsibilities, or deal with the pain of losing a familiar role such as being a parent. Time and again a communication barrier appears as a result of pressure within the family to protect one another by not showing feelings. The social worker can work with the family to try and remove barriers by giving reassurance and helping the family to accept strong, often conflicting feelings such as anger, guilt, sadness, and regret. Some people find it easier to begin to express difficult feelings if someone medical is there to represent security and control of symptoms. In these situations, joint interviews by the social worker and nurse or doctor can be helpful.

As a non-medical person, the social worker can act as a bridge between patients, families, and staff by helping the patient and the family to identify important questions and direct these to the appropriate staff members.

The social worker also has a practical input to make in such issues as discharge planning and locating appropriate financial and other resources in the community. Social workers employed by the local authority have certain statutory responsibilities not only to the elderly, sick, and handicapped, but also a wider mandate to children and the mentally ill. The social worker on the support team may carry out responsibilities such as arranging emergency foster care for the baby of a dying single parent, or helping to assess a patient for compulsory care under the Mental Health Act.

Responsibility for co-ordinating bereavement follow-up work may also be given to the social worker on the support team. This may involve group work or individual contact. The social worker can help the relative explore and express thoughts and feelings which are difficult and painful. By listening, understanding, responding without being judgemental, and being able to tolerate unpleasant feelings, the social worker can prevent the bereaved from feeling overwhelmed, and hopefully help them to move towards adjustment. Bereavement visits can encourage family's questions about the death. These visits can be used to assess whether the person is at risk and requiring more specialist help. All of this demonstrates that the family is still important, and provides an opportunity to say goodbye if contact is to end.

The social worker is often in a position to stand back from the medical and nursing aspects of patient care and help the team acknowledge that they do work as a team. The social worker is able to monitor feelings within the team. The principles of counselling can be used within the team to precipitate sharing and caring, and this ensures that no individual is overwhelmed with responsibility.

Encouraging and accepting the social worker as part of the multi-disciplinary team can lead to a more creative service for the patients, as well as providing for the individual and collective needs of the team.

3.5. CHAPLAIN

Chaplains are usually available in hospitals to attend to the spiritual and appropriate other needs of the patients. Their involvement with support teams is more variable. Some chaplains prefer to be contacted when a need arises and do not meet with the team otherwise. On the other hand, the chaplain may play a very significant role in the functioning of the support team. He or she may be an important source of support and encouragement to the other team members, and may even be lookd upon as a confidant. On occasion, the chaplain will facilitate examination and resolution of team conflict and stress, thereby improving how the team functions.

The degree of involvement with the patients is also variable. Some chaplains have difficulty in dealing with dying patients and their families. Their involvement will often be limited to basic religious requirements such as administering the sacrament of the sick, or reading the prayers for the dying. In the past, there was significantly less training available at theological seminaries on the needs of the terminally ill and their families. More recently, this has tended to change and many chaplains enjoy opportunities for counselling patients and families, and participating in bereavement follow-up. Some chaplains willingly sit with dying patients if the relatives are not available, and go to the home of relatives to break bad news rather than allow them to be phoned.

The chaplain may also fulfil a consumer advocate role. It can be easier for a relative outsider to the medical and nursing professions to consider the patient and family perspective, thereby maintaining a more balanced, ethical approach.

The teaching role can be a very important one, particularly if theological students visit the support team. In addition, chaplains may play an active part in teaching medical, nursing, social work, and other students.

It is very important for the support team to build relationships with non-Christian spiritual advisers, particularly when minority groups are present within the district. Support team members should take time to learn about the cultural and religious needs from these advisers and elders. The support team can help to ensure that their comments are disseminated throughout the wards by means of informal teaching sessions and written documents, such as the *Care of the Dying* pamphlet published by the City and Hackney Health Authority.

3.6. SECRETARY

For maximum efficiency, support teams depend on access to secretarial skills. Smaller teams may have to rely on using secretaries within the general hospital typing pool. This can be very difficult because of staff shortages and the pre-existing volume of work. Relating to one secretary in the pool will be more effective.

Ideally, a secretary should be part of the team. This often has to be on a part-time basis, with only a small number of large multi-disciplinary teams being able to maintain a full-time secretary. A secretary may be a volunteer but this can prove difficult because of the unpredictable nature of the work load.

The secretary will be responsible for maintaining team correspondence, files, and records of team statistics such as number of patients visited. It is also helpful for primary teams and relatives to have the more personal touch of a secretary answering the telephone in the team office, rather than an impersonal, potentially intimidating answerphone. Referrals may be received by the secretary who can discuss the situation and contact a team member if necessary.

The maturity and experience of the secretary will determine whether any other responsibilities may be taken on. It is possible for the secretary to fulfil a counselling role when relatives phone or visit the team. Occasionally, the secretary may visit patients or families at home and in the hospital.

3.7. VOLUNTEERS

The use of volunteers is an integral part of the modern hospice movement. It is doubtful if many hospices would survive were this not the case—certainly the quality of care would deteriorate. Health professionals are not entirely unfamiliar with the presence and contribution of volunteers because many hospitals have a hospital voluntary organization. Within support teams, some professionals will contribute their skills on a voluntary basis. Few teams make use of volunteers in any other capacity.

However, it is possible for support teams to recruit and use volunteers, as demonstrated by the St Thomas's Hospital Terminal Care Support Team (Bates 1985). The use of volunteers requires good organization, including study days and training courses on caring for the terminally ill.

Some of the occasions when volunteers have been used by support teams include the following:

1. Providing transport for patients to clinics and day care facilities. The volunteer driver will frequently find themselves being party to

patient's fears and anxieties as they travel to and from the clinic. This information may help the team to improve their understanding of the patient. It is important to clarify hospital policies about the legal and insurance requirements for transporting patients in private vehicles.

2. Secretarial help.

3. Child minding. Having someone to look after a patient's children may facilitate trips to clinic or allow the patient to have some free time.

4. Shopping.

5. Bereavement follow-up.

6. Sitting with patients to allow family to have a break from care.

7. Sitting with patients at night. This can take place at home and in the hospital. Patients without family, or whose family cannot visit, can be greatly comforted. The anxiety and guilt that the nursing staff feel about not being able to sit with dying patients will also be relieved.

8. Visiting patients at home or in hospital if there are no family or friends. The ability of some people to befriend patients who are socially isolated will also contribute to better patient care and lower stress levels. Patients may well share some problems that they would not otherwise be able to. It is important to recognize that some patients wish to retain their life-long reclusive tendency and will resent the intrusion of 'lay persons' as much as health professionals.

9. Participating in day care. Volunteers can provide expertise in activities such as art and craft in addition to preparing refreshments and befriending patients.

3.8. RECRUITING TEAM MEMBERS

Once it has been established what staff are needed, and what each job entails, the process of recruitment can begin. It is important to advertise widely; Mount (1980) provides guidelines for writing and placing advertisements, and screening applicants by telephone. His article emphasizes the importance of not trying to 'sell' the post. The stress of working on a terminal care support team must not be underplayed or covered up.

The formal interview is a crucial time for assessing potential applicants. Mount's paper also provides recommendations on how to conduct interviews for personnel applying for work in the field of terminal care. He notes that it is important to build up a picture of the whole person, setting aside first impressions based on dress and mannerisms. An examination of past work habits should seek to establish whether the candidate can set and achieve realistic goals (LaGrand 1980). Past education and training should routinely be scrutinized. Enquiring about family background, financial status, and social situation should elicit past and current stressful events, particularly previous deaths and other major losses. The responses to these situations will be important predictors of response to the stress of the new role. A recent bereavement or other significant stress may render the candidate unsuitable, albeit temporarily. The past health record may provide further information about reactions to past stress; general physical condition must be satisfactory, particularly if practical care is required.

The interview, coupled with discussions with previous employers, should evaluate character traits such as stability, ability to relate well, perseverance, and leadership ability. These traits must be considered alongside the candidate's emotional maturity. Flippancy, poor self-control, dependence, inability to accept responsibility, and show-off tendencies will seriously detract from a person's ability to perform in a supportive, advisory role. Conversely, people with the strong sense of self, necessary to withstand the stresses of this field, may be less able to function in a team.

It is important to understand why the candidate applied for the post. Mary Vachon (1978, 1987) has elucidated the motives of people working in terminal care. Some people drift into applying for the job because of the convenience of hours or such like; you should be extremely tentative about employing them. Others seek to master the difficulties of illness and dying, or master colleagues by becoming an advisor. They will find bitter defeat when their advice is not taken. They must be able to accept a certain unremediable incidence of failure, and yet continue with the gentle, indirect approach that is essential for any success at all. A sense of 'calling' may imply a stable philosophy and strong background support but can lead to a crusading approach that overpowers patient, family, and staff. Another double-edged motive is the desire to disprove past failures and demonstrate skills which the candidate perceives have been suppressed in previous jobs.

The final decision about who is appointed should be made on the basis of which applicant has the closest match between capabilities, personality characteristics, and idiosyncracies; and the job requirements. Some degree of compromise will be inevitable. It is vital that the selection committee do not feel compelled to appoint if the degree of mismatch is too great. The apparent needs of the terminally ill and the ready availability of charity money to start a service will conspire to generate a sense of urgency. This urgency must be resisted if a suitable candidate is not forthcoming. Hasty, stop-gap appointments can seriously jeopardize the mental and physical well-being of the person appointed, quite apart from the quality and credibility of the service.

The appointment of team members is merely a prelude to the next difficult phase. The individuals must begin providing a service, and weld themselves into team. The next chapters examine how the team members interact with patients, families, other staff, and with each other.

4

Team dynamics

4.1. INTRODUCTION

Support care teams are subject to a number of forces, from both within the team and without. Team members frequently feel like objects being pulled in several directions by a number of different parties. Only by working together as a team, supporting and encouraging one another, learning from each other, will the necessary sense of competence and self-confidence be maintained to provide a creative service, enjoy the work, and grow as individuals.

The number of support care teams in the U.K. has increased from two in 1977 to thirty-eight listed in the 1989 hospice directory provided by St Christopher's Hospice. However, five teams have collapsed during that time—testimony to the potentially destructive strength of these forces. If a team is to survive, destabilizing forces must be reduced and balanced by forces which maintain the integrity of the individuals and the team.

This chapter seeks to examine the dynamics of support care teams. Broadly speaking, the forces acting on a team derive from influences outside the team (environmental factors), and from how the team functions within itself (internal factors). We have chosen to weave these factors into a sequence, commencing with establishing goals at the inception of the team and then working with patients and families. This leads on to the importance of the decision-making process and advisory role. The impact of other health professionals, medical and nursing students, and visitors to the team is reviewed. The chapter concludes with an examination of how to recognize and manage conflict within the team, and fulfil the educational needs of team members.

4.2. TEAM GOALS

'Team' is more than a descriptive term for a collection of individuals with a common geographic base. Beckhard (1974) recognized that a team is a functional dynamic entity—'a group with a specific task or tasks, the accomplishment of which requires the interdependent and collaborative efforts of its members'.

The cohesion of a support team will require the members to agree upon, and be committed to, a common set of goals which address the needs of patients, families, and staff. The goals outlined in Chapter 1 were based on these needs, and therefore bear repeating.

1. To assist in the relief of distressing symptoms and to give emotional, social, and spiritual support to terminally ill patients.
2. To provide counselling and support to relatives.
3. To provide support and advice to the staff caring for these patients.
4. To take part in education programmes on a multi-disciplinary basis.

These goals emphasize a broad approach encompassing psychosocial and medical needs. They maintain the focus on the importance of a multi-disciplinary approach. Medical problems cannot then become the priority, as is characteristic of the traditional medical model.

Goals need to be adapted to the specific situation, taking into account the other services that exist in the hospital or district. If a home care service is not available, the team may wish to extend their role outside the hospital. The care of relatives can include a bereavement service. The teaching role will depend on the educational input of the local hospice, and hospital in-service education.

Setting limits

Specific limits may be built into team objectives. Patients with benign pain may be referred, particularly if there is no pain clinic, and the team should have a consistent policy for these patients. Several teams restrict their services to terminally ill cancer patients. Supportive care can be provided continuously throughout a terminal illness, or limited to a single review with the primary team responsible for follow-up. Ongoing care occasionally requires specific limits about the length of involvement, particularly when patients are undergoing curative treatment or have early disease. Geographic boundaries need to be clearly defined with teams that border on or work within the district.

Measuring goals

Goals need to be attainable, and team members should be able to appreciate when they have achieved them. Otherwise, uncertainty and dissatisfaction will result. Many symptoms, particularly pain, can be controlled; the positive feedback from patients and families provides a measure of success as well as a powerful reward. The decreased number of complaints about the care of the dying is readily appreciated by the hospital administration (Hockley *et al.* 1988). Feedback from other staff may be less forthcoming initially, but will come to constitute another important source of job satisfaction.

It is important to keep records and compile statistics about workloads. Demonstrating an increase in the number of referrals, visits, home deaths, teaching commitments, and transfers to hospice may help to consolidate the team's position.

Many teams have become involved in projects which formally evaluate their performance. This stems from a need to audit effectiveness, and identify areas of care that need to be improved. Scales for rating severity of symptoms during chemotherapy may be readily adapted. Several workers (MacAdam and Smith 1987; Kristjanson 1986) are developing easily administered questionnaires which also assess emotional, social, and spiritual needs of patients and families. Where possible, 'quality of life' should be assessed by the persons affected, rather than the support team (Slevin *et al.* 1988).

4.3. PATIENT REFERRALS TO THE TEAM

Adopting the aforementioned aims will involve contact with patients and families. From the outset, it is important to define the way in which referrals are made to the team. Nurses frequently want to refer patients; they are more likely to be aware of problems. But the consultant must give permission for the support team to be involved. Confirmation of consent must be given before the patient is seen, particularly if the primary team has never used the support team before. This will relieve doubts and criticisms about 'empire building'. Failure to observe this principle will lead to embarrassment and stress, even jeopardize the future use of the team.

The profile that the support team maintains on the wards will influence the number and stage of referrals. Informal visits to wards will frequently prompt staff to seek advice or refer a patient sooner. This reduces the stress of late referrals but increases the workload. Conversely, home care will limit the time available to spend on the wards. If a team is predominantly involved in community visits, patients may only be referred when they are about to be discharged.

4.4. HOW SUPPORT TEAMS RELATE TO PATIENTS AND FAMILIES

Initial visits

The support team will need to plan how patient referrals will be reviewed. The initial interview is best conducted by someone with experience in symptom control; approximately two thirds of the authors' patients are referred because of distressing symptoms. It is not unusual to find that many patients referred for psychosocial support also have unrecognized distressing symptoms. The nurse specialists usually see most of the patients, particularly when the doctor works part-time. The doctor will then follow-up the initial visit if necessary.

The person with the lowest case load will often see a new referral; some negotiation between team members may be required. Occasionally, circumstances will favour a specific member reviewing a new patient. It may be more appropriate for the doctor on the team to see patients with non-malignant conditions. A support team member who has a good relationship with a difficult primary team should be considered for new referrals from that team. Some primary teams will prefer to relate to nurses, others are less threatened by a doctor.

It may be preferred that the person who conducts the initial interview becomes the key-worker for that patient. This provides a continuity that contrasts with the frequent changes in other staff. The nurses and doctors of the primary team find it helpful to know which member of the support team to contact in the event of new problems. Occasional contacts with other members of the support team minimize disruption of care when the key-worker is sick or on holiday. If the social worker becomes the key-worker, a nurse will often maintain a minimal contact; regular joint visits by the nurse and social worker may be appropriate with some patients.

Continuity of care is particularly important when a patient is transferred to another ward or primary team. The key-worker can focus the attention of the new staff by telling them what has already happened to the patient and family. If more than one primary team is involved, it can be useful to identify for the patient and family which consultant is dealing with which problem.

Some teams prefer new patients to be seen by two team members. This has the advantage of providing different insights and perspectives about the patient's problems. There are also fewer difficulties when team members are absent for any reason.

It is very important to establish limits on the number of referrals that are accepted. This will minimize role overload. Although it is difficult to generalize, an average of two new hospital referrals a week per full-time equivalent nurse or doctor on the team is probably a comfortable maximum. The nature of the case load will influence this; three or four difficult symptom control or psychosocial problems may be more taxing than ten 'straightforward' patients. Lunt and Hillier (1981) considered sixty new patients a year per nurse to be an ideal for home care.

Follow-up visits

The frequency of follow-up visits should be carefully considered. Patients can be seen every day but this may prove counter-productive. Doctors and nurses may reduce contact with the patient because they see the support team involved so frequently. Most teams review in-patients two or three times a week on average. Patients with very distressing symptoms will be an exception; they may require review several times a day until their symptoms are brought under control.

The number of visits should enable sufficient time for the interviews to be relaxed and unhurried. It may take several visits before the patient opens up about their deepest fears and anxieties; they are accustomed to accepting symptoms and worries rather than interrupting the nurses' busy schedule or the doctors' hurried routine visits. One must show there is the time and the strength to accept their fears, anger, and grief.

Commitment to the patient is demonstrated by being reliable, visiting as promised and following up any requests from the patient. Facilitating consultations by visiting the surgeon or physician, or finding results of tests by visiting the laboratory or X-ray department, may also help.

Given that a higher percentage of patients die within weeks of referral it is important to recognize the value of even one family session. Asking to see a family together is a statement of the team's concern and the family's importance. A family meeting can generate confidence in a situation that feels out of control. The session can be used to encourage family strengths, help them to identify and include other supports around them such as church, and address what is reasonable for them to expect of each other.

Out-patient contact

A few teams provide an out-patient clinic. These clinics are kept small, with only about six or seven patients attending one afternoon a week. This allows time for the patients and relatives to talk about their problems.

When patients attend other clinics, the team member tries to see them before they go in to see the consultant or senior registrar. This helps to minimize the effect of delays. Accompanying the patient facilitates the interview and increases contact with the doctors. If transport to clinic is a difficulty, a volunteer scheme or a taxi service should be arranged rather than have the patient wait for hospital transport.

Patients value regular telephone contacts. If this is not maintained, patients may be re-admitted without the team knowing for several days. One solution could be the use of a sticker on the hospital notes which requests that the primary team contact the support team immediately the patient is admitted.

Bereavement follow-up

Following the death of a patient, the relatives should be seen when they collect the death certificate and the patient's belongings. Contact can be maintained with the family by telephone call, letter, or follow-up visit.

Bereavement follow-up can be difficult because of the time required. For this reason, teams operating a follow-up programme usually have a social worker; few teams are fortunate enough to have the services of a bereavement co-ordinator and volunteers. Some teams rely on referring to existing community or hospice services.

Regular meetings for the bereaved can open discussion about coping with loss, adjusting to isolation, dealing with the disposal of loved one's clothing, family support, and gradually becoming independent. Casualty and intensive care services may refer bereaved families. Groups can be organized on a continuous monthly basis or limited to a specific number of weekly sessions. The intensity of this type of group counselling combined with the day-to-day stresses experienced by the team can be very taxing.

4.5. THE STRESS OF WORKING WITH PATIENTS AND FAMILIES

Working with terminally ill patients and families can be very stressful. Extremes of emotional reaction, especially anger and aggression, are very difficult to cope with. These emotions are predictable responses to the devastating impact of a terminal illness. But that does not make these reactions any easier to deal with, particularly when they are directed at the members of the team rather than the cancer.

Patients come from a variety of cultural, religious, and social backgrounds. This can make interpretation of their reactions more difficult, particularly if language barriers exist. Translators do not necessarily help. They often share the same powerful cultural forces which inhibit dealing with distressing information. In addition, truth-telling may not be the norm in other cultures and considerable distress will occur if this is not respected.

Patients referred at the end of their illness are a source of considerable stress. Distressing symptoms make the patient seem alert but their relief will restore the patient to a state of dying peacefully. However, the distress of the relatives will increase because they perceive the change in the patient to be an attempt to accelerate death. The difficulty in dealing with these reactions is compounded by the brief time available, and the frustration of thinking the task would have been easier if the patient had been referred earlier.

Situations in which families, but not patients, are told the diagnosis are becoming less frequent as the trend towards truth-telling increases. However, it is very stressful when families and the primary team pursue a strategy of collusion, particularly if the patient is distressed by his ignorance and is asking for more information. Breaking a collusion will provoke an intense reaction from the relatives. Patience, diplomacy, and support are required to resolve these problems.

Families can become angry with the support team for helping them to accept the impending loss of a loved one, only to have the patient recover for a time. A similar situation may occur when patients and families attempt to patch up chronic problems. Long-standing disagreements may be set aside in an effort to make the patient's final days as enjoyable as possible. If the illness becomes protracted, old tensions may be refuelled by the frustration of the patient not dying, and then directed at the team.

It is not always possible to control pain, breathlessness, and other distressing symptoms, particularly when medications cause unacceptable side-effects. Some patients and families will be angry at this failure, using the opportunity to displace anger about the illness. A sense of impotence and failure may be further heightened if it is felt that one's 'expert' status has been weakened in the minds of the other staff.

Patient denial can also cause a sense of failure if there is a strong feeling that a patient should accept their illness. But there can be major difficulties if the team perseveres in trying to undermine it. Quite apart from the extra suffering experienced by the patient, they can become angry and reject the team. This may inhibit doctors from referring other patients.

Patients who exercise denial may go so far as to discontinue all medications, including pain killers, in the belief that they are cured. It is even more common to find patients who want any attempt at curative treatment. Anything less is totally unacceptable. Their desperation may lead them to press for radical treatments such as aggressive chemotherapy, even if there has been no previous response, or disfiguring surgery. Some team members will be repulsed by these choices, which brings us to the next important element of team dynamics: what is the basis for making decisions about, and therefore advising on, patient management?

4.6. THE ETHICAL BASIS FOR MAKING DECISIONS

One only has to consider the number of people involved in deciding what is best for the patient—the patient himself, the family, the nurses, and doctors—to realize the potential for conflict. Into this situation comes the support team, a 'problem solving, decision making mechanism' (Beckhard 1974) which can only advise. It is essential for team members to have a sound understanding of the ethical basis for making decisions. Otherwise, it will simply be adding yet another opinion to the conflict.

Current medical ethics emphasize that patient choice is central to the decision-making process (Beauchamp and Childress 1983). Patients are autonomous individuals capable of choosing their treatment and care. Making decisions for a patient, without their consent, is only ethically justified when they are not capable of understanding or making decisions. One cannot label patients as incompetent just because they are unco-operative. However, patients who are confused, psychotic, demented, or comatose when dying may be unable to comprehend or decide. Severely depressed patients may also have difficulty making decisions. Any wishes expressed by patients before they become incompetent must be taken into account.

When asked to see a patient, one must define what the problems are, and who has which problem. Team members will soon learn that hospital staff may present problems such as pain when it is actually the staff who are having difficulty coping with the patient.

The first task is to find out what the patient perceives to be problems. One should begin with open-ended questions such as 'how is it affecting you?' When the patient has described his needs, direct questions may be used but try not to put words in the patient's mouth. Acknowledging a patient's wish to ignore or deny distress from symptoms must be respected. Demonstrating a commitment to the patient's wishes will allow him to maintain a sense of control. It also saves the team and the patient from the stress of dealing with issues which are not important.

It is important to interview relatives and staff. Valuable information can be obtained about symptoms or psychosocial difficulties, and it may be possible to find out about past experiences of cancer which are dictating what the patient presents or how he interprets information (Billings 1985). This information should not override the patient's perceptions, but it may subsequently facilitate expression of fears about symptoms or treatment.

Finding out what the patient wants is vital, but the principle of comparative justice recognizes that the choice of the patient is not absolute. The needs of the family are important, particularly if they are going to care for the patient at home. The staff have commitments to other patients, and patient choice may conflict with their values, such as a patient who asks the staff to carry out euthanasia. Balancing these other priorities with the choice of the patient can be very difficult.

As problems are defined, possible solutions can be devised. These strategies should then be presented to the patient. He should be given the opportunity to ask questions about potential risks and benefits, and alternative treatments. The patient should also have time to deliberate, if he wants it. The team should then endeavour to support the patient's choice, but should remember that patients may change their views with time.

Patients find it difficult to participate in discussions about strategies for their needs because they fear that the doctors may take umbrage. It is little wonder when reading the St Bartholomew's Hospital rules published *circa* 1900. Rule 7 stated that 'every patient must strictly obey the Directions of the Physician or Surgeon under whose care he or she may be placed'. Compliance was ensured by the concluding rule; 'any patient acting contrary to the foregoing Rules will be reported by the Sister of the Ward to the Steward or Matron, and by them to the Treasurer: such Patient will be admonished or discharged'.

It is also important to recognize that, while a patient may initially agree to a particular treatment, later non-compliance should be recognized as a choice, not as being 'unreasonable' or 'ungrateful'.

Failure to co-operate is one of the few tactics that patients have for maintaining control.

Maintaining the precedence of informed patient choice can be very difficult, particularly if a philosophy of palliative care is exercised which minimizes active treatment. In the hospital situation, one does not want the patient becoming a battle field between the 'care oriented' team member and the 'cure oriented' hospital teams. However, patients may well accept a considerable degree of risk from treatments which hold minimal chances of benefit.

In practice, it may be extremely difficult to present alternatives to the patient, particularly when the primary team is not willing or likely to fulfil the patient's choice. In some circumstances, a patient will be referred because he wants more treatment than the primary team are prepared to give. The expectation of the primary team will be that the team member will persuade the patient to accept their decision. One must never lose sight of these expectations but, for the patient's sake, try to look beyond them. How adjustments are made to conform with these expectations and patient choice is covered later in this chapter.

A course of action may need to be confirmed immediately with the patient, particularly when symptoms are causing distress. Previous experience in palliative care will usually enable advice to be provided at the time. One should feel comfortable about asking another team member to come to the ward if there is uncertainty about the situation.

The team should have a forum for discussing strategies for needs that are not so immediate. The experience of the Charing Cross Hospital Support Team illustrates the importance of the team members meeting together to make decisions. Failure in communication was one of the factors cited in the collapse of the team (Herxheimer *et al.* 1985). Most teams have at least one regular meeting a week to discuss new patients and ongoing problems. Team meetings concentrate the variety of disciplines and individual talents of the members. Problem definition is greatly enhanced and several perspectives can be generated.

Some problems lie outside the collective experience of the team, such as a difficult anxiety state or specific symptom problem that may need psychiatric consultation or a more expert opinion. One should recognize these limitations and not be afraid to seek the advice of others. It can be particularly helpful to discuss problems with the local hospice, or a major teaching hospice. Pride may make this process difficult within the hospital. However, any fear of appearing incompetent must be set aside. Other staff may provide useful suggestions, and they appreciate the chance to be involved in this way. They will actually feel more comfortable about using the team. Even if the advice is not totally appropriate, the communication between both parties can only help raise the profile of the terminally ill. It is important to remember those professionals who are prepared to consider the plight of the terminally ill and who may therefore be of help in the future.

Even though a problem remains unsolved and all possibilities exhausted, patients, family, and staff will still benefit from ongoing support. However, a key-worker is likely to become worn down by frustration and disappointment, especially if the illness is protracted. Other team members should share the load by becoming involved with the patient and family.

4.7. COMMUNICATING ADVICE TO THE PRIMARY TEAM

We have now come to the point where, having discussed the issues with the patient and the rest of the support team, you must advise the primary team. It will often seem easier to pass on the advice and run, but a commitment to patient choice will usually compel team members to persist as the patient's advocate when advice is ignored. This is when the harsh reality of the advisory role becomes apparent.

Doctors often have difficulty asking for and accepting advice. This is partly due to a lack of training in teamwork. It also represents a conservative tendency designed to protect the patient from harmful treatments. The support team will be treated like a new drug: something with therapeutic potential, yet to be realized, and side-effects which must be balanced against that potential. Unfortunately, primary teams often presume that the side-effects will include increased patient anxiety and fear, even before the team begins to function. These presumptions are inherent in the way some support teams have initially been labelled 'death squad' or 'angel of death'.

When first approaching a primary team, doctors may well rigorously question advice given, particularly in teaching hospitals. If you have trained in hospice, where there is mutual agreement about treatments, it will be very disconcerting when evidence to support the advice is demanded, esoteric side-effects are pointed out, or alternatives are put forward. The interrogation is not designed to attack the team member but to serve as a reminder of who is responsible for the patient. In this situation, one must resist manufacturing answers for the sake of preserving credibility; a simple 'I do not know' will have more impact. Knowing that the advice represents a team consensus will give you more confidence, as will the company of another team member.

With time, most doctors will relax this tactic as they become aware of the efficacy and non-judgemental nature of the support team. However, further challenges should be expected whenever the registrars or housestaff change-over.

As the support team becomes accepted, advice about minor alterations in medications or treatment will require only informal discussion with medical or nursing staff. Comments can be written in the medical or nursing notes, but if prompt action is needed the staff concerned should be spoken to.

It is very tempting for the support team doctor to expedite treatment changes by writing prescriptions. However, this will rob the doctors of a valuable teaching experience. It may take more time and effort to contact the junior doctor but greater dividends will result. Rarely, the doctor on the support team may need to prescribe because the other staff are occupied in the operating theatre or 'on-call'.

Advice which is radically different from the thinking of the primary team should be discussed with the consultant or registrar. The more formal setting of a ward round can be used, but this may be very daunting. Many consultants or senior registrars can be approached before a ward round; they may be more receptive to new suggestions which might otherwise be rejected in the more public setting of the round.

Attending ward rounds or multi-disciplinary meetings, even when not involved with patients, can be a very valuable way to break down suspicions that the team feel they are somehow better than everyone else. Informal meetings over lunch can also have a similar effect. After a while, questions about terminal care will be asked, and patients may be referred.

4.8. WHAT TO DO IF ADVICE IS DECLINED

If the advice of the support team member is declined or ignored, the suffering of the patient and family will continue unabated. This can be extremely upsetting, particularly when one contrasts the suffering with the dignity of patients in hospice. It is even more disturbing when a patient's anguish is increased by the team protesting about the failure to act on recommendations. Challenging the primary team may generate anger which is displaced onto the patient, either directly or by avoiding the patient.

The realization of the responsibility, even in a small part, for an increase in suffering can be very depressing and demoralizing. These emotions may exceed feelings of anger and frustration, and may lead to a sense of only having a token role with nothing to balance the loss of familiar skills, such as performing 'hands on' care, entailed in joining the support team.

When confronted by this situation, a decision must be made whether to persevere. It may be better to opt out altogether, particularly if feelings are so intense that they are likely to spill over when talking to the primary team. Support from colleagues will be necessary whether you stay involved or not.

If remaining involved, try and determine why the primary team have not accepted the advice. Doctors usually make decisions based on what they feel is 'best for the patient'. This paternalistic approach is supported by case law. The right of the patient to make an informed judgement about his treatment has only received minority support from the legal profession (Sidaway vs. Bethlem Board of Governors). Currently, 'a doctor is not guilty of negligence if he has acted in accordance with a practice accepted as proper by a reasonable body of medical men skilled in that particular art' (Bolam vs. Friern Hospital Management Committee). This principle, the Bolam principle, encompasses 'the practice of saying very little and waiting for questions from the patient' (Bolam vs. Friern Hospital Management Committee).

These medico-legal considerations mean that some doctors must be assured that advice is based on 'accepted' practice before they will accept it. It is possible to reverse some decisions if evidence from the literature can be provided. This underlines the importance of knowing about key papers on palliative care.

The paternalistic basis for decisions frequently reflects a genuine altruistic response. Appealing to such charitable motives by presenting more personal details can help the doctors realize they are dealing with a person rather than a 'case'. Less commonly, paternalism signifies a rigid authoritarian position. This attitude is the most difficult to influence. It is in this situation that the development of alternative strategies can facilitate a compromise. In addition, the support team member who is best suited to liaising with the primary team should be carefully selected. A member of the nursing staff may be less threatening; on other occasions the credibility of the team doctor may be more important.

If advice is rejected, wait and see if the suggestion is actually taken up. It is surprising how often the primary team will subsequently change the patient's treatment. The change is often a compromise in favour of suggestions made rather than wholehearted acceptance. For example, the primary team might increase subtherapeutic parenteral doses of pethidine (meperidine) rather than change to oral morphine. Any efforts in the right direction should be praised. The best way to do this is to tell the primary team that the patient is grateful for the improvement in pain control. One can then mention that the patient is still in pain and ask if they would consider increasing the dose and frequency.

Negotiating around what is familiar to the primary team will usually produce a more satisfactory result for the patient, but the reader will appreciate that the compromise may involve the patient continuing to experience distressing symptoms. Such a compromise is never ideal. The stress may be lessened by recognizing some progress will have been made, if not for the patient in question, then for subsequent patients. It is the degree of compromise that is acceptable which will determine whether it is possible to continue to work in this advisory role.

The alternative of not talking directly to the primary team may seem very attractive. It is less stressful to write in the hospital notes or to use the nursing or junior medical staff to pass on your comments, but these staff then experience the stress of being caught in the middle of the support team and consultant or senior registrar. Progress is even less likely when this happens.

4.9. OTHER STRESSES ON A SUPPORT TEAM

Hospital services

The expectations that primary teams have about the support team may produce considerable strain. Many consultants hold the view that the function of a support team is to transfer patients to a hospice. This view arises from the clinician's dilemma when there are delays in potentially curable cancer operations because of a shortage of acute beds. The pressure can seriously undermine the ability of the support team to establish a rapport with the patient. Many patients will welcome the opportunity to go to a hospice; it is important to recognize these patients quickly so that they can benefit from early transfer. Others will be extremely upset. The consultant may respond to a special plea, especially if the nursing staff on the ward support the patient's choice. On the few occasions when a reversal of the decision has not been obtained, the patients have died very quickly after transfer.

Problems also arise when terminally ill patients or relatives request readmission to hospital. If the patient is well known to a ward, this can often be arranged easily. However, lack of acute beds on the ward may require the patient to be admitted via Accident and Emergency or a 'holding' ward. The admitting team or A&E staff may be angry and obstruct the admission, even discharge the patient. Terminally ill patients are not yet considered emergencies, even though their weakened state magnifies the intensity of symptoms and the distress of relatives. Where possible, the nurse specialist should arrange to meet the patient and family in Casualty and accompany them to the ward.

Community services

Support teams working with general practitioners and district nurses will find that the advisory role can be just as difficult in the community. Many problems, such as reluctance to use strong opioid pain-killers, are common to the hospital setting. These problems are compounded by the suspicions about hospital-based staff encroaching into the community. Diplomacy and flexibility are just as necessary for overcoming these fears.

Hospices and home care teams

Hospital support teams will frequently need to liaise with local hospices and home care teams. In general, liaison should not be difficult particularly if you have allayed misapprehensions during the planning of the team. The team should be aware of the specific limitations in other services. Smaller units will have difficulty taking in patients for more than two or three weeks respite care. Some hospices and home care services will only take patients in the final weeks of life. Problems may arise with 'shared care'; working with patients who are still receiving active treatment may be unacceptable to some hospices.

Lectures and visitors

The number of lectures and visitors to the team can contribute to inter-role conflict. Many people find the exhilaration and feedback of teaching provides an important source of job satisfaction, but limits need to be established or the service to patients may suffer. More than one or two visitors per week can be very disruptive to the integrity of the team.

4.10. RELATING TOGETHER AS A TEAM

Team meeting

Team meetings provide the forum for making decisions. The traditional medical model requires the doctor to be the team leader. This has important consequences for the dynamics of meetings. 'Leader' is synonymous with 'decision-maker', particularly when the doctor is accountable for wrong decisions. Some teams are comfortable with this model. Decisions will tend to be made on a unilateral basis, or by default if non-medical staff feel compelled to stay quiet when they disagree. Occasionally, this will lead to team members lacking the commitment necessary to action decisions.

The nurse specialists will be less confident about making urgent decisions at the patient's bedside. Difficulties will arise if the doctor is frequently unavailable, or new members of the team want a greater role in the decision-making process.

Some teams may wish to develop beyond the traditional model. It takes a lot of effort to break down conventional, status-determined patterns. The observations and opinions of all team members must be affirmed and valued; too often, praise is considered inappropriate for professionals doing a job. Senior team members, particularly the doctor, should be sure that they do not cut short information from other team members. The aim should be to build up a variety of strategies based on collective experience, rather than the 'right' way. This model has more flexibility for dealing with non-medical problems.

Team reviews

Team reviews serve as a forum for stress and conflict management. They should be a regular feature of team life, remaining distinct from the daily and weekly team meetings described previously. A few teams make use of an 'objective' facilitator, someone who is not normally part of the team. One team uses a management group comprising a community physician, director of nursing services, senior hospital social worker, and chaplain to handle problems and personal matters. The group meets regularly with the support team.

For reviews to be useful, team members should become familiar with the causes and manifestations of stress within themselves and their colleagues. Vachon's reviews (1986, 1987) of stress and the care of the dying are highly recommended. Some stress may be valuable as a stimulus to better performance. However, when coping mechanisms are overwhelmed, physical symptoms, such as fatigue, weight change, and sleep disturbance, may develop. Feelings of guilt, anger, irritability, and frustration will generate conflicts in the team and the individual's private life. If stress continues unabated, confidence will be undermined, and the team member will experience difficulty in making or implementing decisions. Eventually, depression and a sense of helplessness will render the individual incapable of continuing in their job.

The initial reviews will often consist of sharing anxieties about being exposed in the advisory role. The stresses inherent in coming to terms with new roles, a new environment, and a new set of colleagues will be submerged by the excitement of embarking on a new venture, and the need to build up a sense of 'being in it together'.

Eventually, the 'honeymoon' excitement that sustains team members during the initial months will wane. The reality of trying to meet the needs of the terminally ill and their carers will become starkly apparent, as will any mismatch between team members' skills and job requirements. The impact of patient deaths will hit home, particularly for team members who have not worked in hospice. Review sessions will be characterized by the venting of frustrations about people outside the team—the 'uncooperative' doctors, 'unfeeling' nurses, and 'ungrateful' patients.

All the while, team members will become increasingly aware of personality and professional conflicts within the team. However, the usual tendency is to suppress disagreement and conflict because there is a need to have the team's approval and the reaffirmation that you are nice and doing the right thing. It is important to recognize that 'chronic niceness'—each team member wearing a smiling mask that hides ill-feelings—may be very important for the functional integrity of the team, at least in the short-term.

There are problems with maintaining a facade. Team members will often sublimate personal conflict into their job. But expending mental energy to maintain an illusion of team cohesion will detract from performing clinical tasks. Tension may also be eased by discussing the situation with others on the team but not with the person who is perceived to be the cause of the conflict. This has the long-term effect of creating factions and further disruption. Releasing tensions after work will disrupt relationships with family and friends.

Taking down the masks of 'chronic niceness' is very difficult. It can take several months before safety within the team permits the exploring of disagreements. Otherwise, criticism will be perceived as a personal attack and will result in loss of self-esteem. It is vital to build up an atmosphere of trust and mutual respect by encouraging team member's skills and strengths. This can be done by sharing personal successes and achievements at a team level, and allowing others to give due praise. This affirmation has to be realistic and not paternalistic.

In a supportive atmosphere, constructive solutions can be explored if job mismatch is exposed. It may be appropriate to alter the job description to better suit the team member or encourage the learning of new skills. In some cases, situations which exacerbate the problem should be avoided; occasionally this may necessitate leaving the job.

Team reviews should also include social gatherings. Pub lunches, picnics, theatre visits, and other outings are relaxing and therapeutic. They stop reviews from becoming sombre and introspective. Team members, particularly new ones, find that it is much easier in these situations to establish relationships with each other, particularly if there are lingering memories about the person who occupied the post previously.

Reviews can also be useful for reflecting on deaths which have been particularly difficult. Attempts should be made to define 'what' and 'who' contributed to the stressful death. If possible, alternative strategies should be defined to prevent similar problems happening again.

4.11. FULFILLING THE EDUCATIONAL NEEDS OF TEAM MEMBERS

Acquiring new knowledge and skills can make the job more rewarding. Quite apart from a sense of personal growth, more satisfaction will be gained from being able to provide a better service. Attending courses and seminars will also help to overcome any sense of isolation. Team members should support one another by encouraging study leave.

The annual conference for hospital support teams held at St Christopher's Hospice provides an excellent chance to share experiences and frustrations, and explore ways of improving the service. An update on symptom control is provided. There are many other conferences on terminal care. The National Society for Cancer Relief, Help the Hospices, special team funds, and other hospital funds are the main sources by which expenses for conferences are met.

It is not sufficient to just increase knowledge when wanting to acquire or refine skills, it is necessary to practise skills in a controlled environment. Counselling is an example of a skill that is very important for support team work; there are a number of workshops which are very useful. Teaching skills can also be improved in workshop seminars. Mention has already been made of working in a hospice to improve practical skills.

It is helpful for team members to research clinical problems which arise, and then present a short summary to the team. The research and presentation should not be time-consuming, nor relate solely to medical problems. It has proved to be a useful way of improving self-confidence. The traditional model of doctor teaching doctors, nurse teaching nurse, and social worker teaching social worker has been broken down. This has further strengthened the team approach.

A sound knowledge of terminal care, and the skills to pass on this knowledge, are essential for the educational role of the support team. This role is examined in more detail in the next chapter. The other ways in which the team supports the doctors, nurses, and other hospital staff are also reviewed.

5

Supporting the professional carers

5.1. INTRODUCTION

The role of the support care team involves helping patients and families to understand, talk about, and perhaps come to terms with the situation they face. The needs of the other hospital staff, particularly the ward nurses and doctors, also have a major impact on the support care team. Many times, team members will find themselves encouraging and supporting these staff. If these needs are not recognized or met, patient care will suffer; nursing and medical staff may continue to distance themselves by ignoring patients or discharging them prematurely.

The task of meeting the needs of other carers is made more difficult by the fact that many health professionals do not like to acknowledge stress. There are good reasons for this: suppression of intense emotions is necessary to maintain a calm demeanour when dealing with emergencies, and staff should not burden patients with fears and anxieties or transfer anger and frustration onto patients. However, these practices have developed into an unwritten code that professionals do not experience stress, or if they do, it is not acceptable to share it with colleagues. Pride plays a part in generating these attitudes—the need to be in control. Past experiences, particularly during training, also strongly support this unwritten code.

Student nurses who enter into training in hospitals have usually never had any experience of coping with death. However, within a matter of weeks they are often caring for patients who die. Some student nurses working on their first ward will be involved with at least 5–10 dying patients. Whitfield (1979) found this was one of their most stressful experiences, and they received little formal teaching to prepare them. The busy nature of acute wards, coupled with the vulnerability of senior staff to the same stresses, frequently results in the student being unable to share these experiences. Crying elicits advice to 'pull yourself together' or an uncomfortable silence emphasizing that such feelings should be kept under control. It is hardly surprising that such experiences may be relived with considerable emotional intensity many years later.

By contrast medical students have much less exposure to the terminally ill. Up to 30 per cent of medical students graduate without ever having had any active participation in the management of dying patients (Ahmedzai 1982). However, considerable stress is generated by the need to acquire the knowledge and skills to pass exams and become a competent doctor (McCue 1982). It is rare for the stress of medical training to be acknowledged beyond a relatively intimate group of peers. Furthermore, medical students frequently develop a toughened exterior to withstand a teaching programme which does not permit the student saying 'I do not know'. As the doctor progresses into postgraduate training, even peer support will tend to subside and the pressure to maintain a smooth unruffled professional appearance will become greater.

The difficulties in recognizing and dealing with stress will frequently cause primary teams to present the patients, families, and even other members of staff as problems for the support team to deal with. The true situation will become apparent to the support team when the patient denies being distressed by or even having the symptoms that were recorded in the referral. Any effort by the support team to shift the focus back to the primary caregivers will be resisted and even resented.

The remainder of this chapter examines the problems that nurses, doctors, and other health professionals experience when looking after the terminally ill. Ways in which support teams can help with these problems have also been identified. Particular attention has been paid to the role of education.

5.2. NURSING STAFF

Patient and family related stresses

The majority of nurses identify their prime aim in caring for the terminally ill as keeping the patient as pain free and comfortable as possible. However, they feel this is only sometimes achieved. Two thirds of nurses find physical symptoms such as pain, breathlessness, and the 'death rattle' are a distressing aspect of caring for the dying (Hockley 1989). Clearly, the support team can reduce this distress by improving patients' symptoms, and educating the nurses about the principles of symptom control.

Unfortunately, nurses find that doctors often prescribe inadequate doses of pain killers or will not respond to patients' needs at all. It becomes easier to avoid patients than to find out they are suffering and be unable to comfort. This will manifest as delays in responding to patient calls and minimal time being spent with patients. If the support team introduces the tools for change, nurses will quickly become more responsive. However, their frustration will be increased if re-education of the doctors is not an integral part of the team's teaching role.

The majority of nurses find performing good nursing care rewarding; it is frequently identified as a major reason for entering the profession. Practical procedures such as dealing with incontinence and vomiting, dressing offensive wounds, and feeding the patient are generally not stressful because they involve caring for the dying patient. However, nurses' anxiety will be increased if a patient refuses a bed bath or other nursing care, particularly if the patient is reclusive and withdrawn. Anxiety will be made worse by feelings of guilt: that the patient is being let down, or that the situation reflects some fault or inability of the nurses. Furthermore, the nurses may feel angry because the patient is 'cheating' them out of physical care.

In this situation, support team members should recognize and respect these feelings. It is equally important to establish the needs of the patient, and why they wish to minimize nursing care. The nursing staff will often be relieved when the support team member relays to them that the patient's decision resulted in the problem rather than the incompetence of the nurses. Respecting the patient's wishes then becomes a positive step which the nurses can use to counter the loss of satisfaction from not carrying out nursing care.

The psychological needs of the patient and family are frequently rated as distressing. Anxious patients, or patients who ask questions about whether they are dying, cause as much stress as uncontrolled physical symptoms. Difficulties arise with young patients; nurses may identify with the patient and young family. Aggressive patients may arouse anger and frustration, and, at the same time, a sense of guilt; negative feelings are not considered part of the professional carers' repertoire.

The stress of caring for relatives reflects, in part, the tendency for families to displace anger and anticipatory grief onto the ward staff. It is compounded by the fact that at least 50 per cent of relatives of dying patients may not see any doctors during their loved one's final admission (Hockley *et al.* 1988). Even when the patient is relatively well, doctors often fail to convey the reasoning behind important treatment decisions. When these decisions are delivered, the patient and family do not ask questions at the time. Subsequently, the nursing staff have to bear the brunt of the patient's and family's criticisms and questions about these decisions.

Other patients in the immediate proximity of a dying patient can cause stress. Neighbouring patients are often not told when a terminally ill patient dies; usually they are simply left to guess. Occasionally they will be given an alternative story such as the deceased having been transferred to another hospital. The other nurses then face the added stress of participating in a collusion and the fear that their relationship with these patients will be undermined.

Support team members may feel more comfortable dealing with these situations, particularly if they have trained in hospice. However, they must resist the tendency to take over from the ward staff. Although stress levels will be reduced, the nurses may not learn by the experience and may become more reliant on the support team. The ward staff, particularly the patient's primary nurse, should be encouraged to accompany the support team member when the patient is seen. Observation, and hopefully participation, will improve nurses' confidence. Future efforts at managing similar psychological problems should be recognized by the support team and given due praise.

There will be difficult situations when symptoms cannot be controlled or psychosocial problems resolved. The support team may feel the same sense of failure and stress as the primary team. It is just as important for the support team to continue supporting the nursing staff as it is to continue visiting the patient and family—the nurses have to continue caring. Even if nothing more is possible, participating in the situation will help the nursing staff. Sharing frustrations and a sense of inadequacy will encourage them to accept their feelings.

The ongoing reduction in the number of nurses working on wards is very stressful. It creates difficulties in carrying out nursing procedures, and contributes to high rates of staff turnover. Student nurses may be left in charge of wards where patients are dying. It can be difficult, particularly after hours, to get senior nurses to come to the ward and check out opiates for patients in pain.

It is usually impossible for the support team to influence those factors—they may make the situation worse. The ward staff can become jealous of the time that the team members have to sit and talk with patients, only then to give advice and not contribute practically. Helping with procedures may be a valuable public relations exercise at such times. It is also helpful to talk about the difficulties that the ward staff are experiencing, even though you may be powerless to bring about any change. The fact that someone is able to appreciate the problems will contrast with the administration who continually expect that standards of care will be maintained despite reduced staff numbers and expertise.

Inter-professional stresses

Nurses often have difficulty participating in administration of treatments, which places them in conflict with the doctors. The difficulty arises because patients often share their deepest fears and anxieties about the medical management with the nurses. At times, nurses misinterpret what the patient is saying—'I do not like treatment' is not the same as 'I do not want treatment'—or they transfer their own fears about treatment. In these situations, it is important for the support team to clarify what the patient is really saying and help the nurses to support the patient's choice. However, when the nurse is in tune with the wishes of the patient, the support team should assist in taking up the issues with the medical team.

When it is not possible to change the attitudes or decisions of the medical team, it will still be important to confirm that the nurses are correct in their interpretation. This helps the nursing staff maintain their sense of personal worth and ability which is so often undermined when their perceptions are at variance with the doctors—'the doctors must be right'. Relaying the doctors' perceptions of the problem and reasoning for decisions, if these can be identified, may be valuable. It is usually very difficult to facilitate a situation of open dialogue.

The continuing turnover of junior house staff may cause problems. They are usually responsible for prescribing the medications necessary for symptom control. Patient care suffers when the doctor is not familiar with using drugs like morphine and nurses have difficulty being confident about details of prescription when asked for guidance. The stress of educating each new set of house staff is made easier by involving the team.

The nursing staff will usually be first to realize the support team's potential for reducing stress. They will often want to refer terminally ill patients before the doctors do. If the team accepts referrals from nurses, there is a risk of exacerbating stress should the doctors take exception.

Informal support

The pre-existing support for nurses working with terminally ill patients varies considerably. It will largely depend on the attitude and expertise of the ward sister and the other senior nurses. The nursing process will often be helpful because the individual nurse is expected to establish close relationships with patients and their families, and then discuss any problems that arise. Support teams must be aware of and strengthen what is available.

Support team members should appear to have time for individual nurses, and be approachable when on the wards. If the ward is not busy, spend short periods of time socializing, perhaps chatting informally over a cup of tea. This will serve to monitor the emotional state of the wards. By getting to know the nurses on the various wards before crises develop, you can identify who might be at risk from stress. It will also help the nurses feel less threatened when you come to discuss difficulties.

After a difficult patient or a number of deaths in rapid succession, it can be helpful to arrange an informal meeting with the nurses. These sessions should not be didactic in nature; all the nurses, including the students, should be encouraged to share their experiences. This provides an opportunity to share feelings about difficulties and to identify topics for further discussion and education. Staff confidence can be restored by acknowledging aspects of care that were done well.

Teaching for nurses

The nurse specialists are often called upon to give formal lectures to student and pupil nurses, auxillary nurses, and trained staff. The number of teaching sessions available varies according to the interest of the nursing tutors.

When planning the content of lectures, it is important to be aware of the needs perceived by the nurses. These needs vary considerably with length of experience and seniority.

Where possible, it is helpful to spend time with the student nurses at each year in their training. Too often the tendency is to cluster the teaching about terminal illness into the initial or final stages of their training with the result that students feel overwhelmed.

The introductory, first, and second year student nurses want preparation for their feelings and reactions to terminal illness and death. Student nurses worry about how to approach dying patients—how to talk to these patients naturally and also how to respond to difficult questions such as 'am I dying?'. They find it helpful to be forewarned about the feelings of being stunned and shocked by death, and to be reassured that these feelings are normal and acceptable. Students feel more comfortable about sharing fears and anxieties in small group sessions rather than lectures. Formal presentations, particularly when accompanied by emotive films, may arouse considerable feeling and reaction that should be given opportunity for release.

By the end of their final year in training, many student nurses can talk to patients without feeling self-conscious. They become more aware of the needs of families and want to learn more about coping with the relatives' questions and emotions. The fundamentals of symptom control can be intermingled with teaching on psychodynamics.

Trained nurses quickly realize they can influence patient management. Junior doctors often approach senior nursing staff for advice about managing pain; this provides the motivation to learn more about the specifics of symptom control.

Lectures on study days, symposia, and post-graduate courses can be used to teach symptom control. Smaller, informal ward meetings are also ideal formats; they can be arranged to coincide with staff overlap in the afternoons. These sessions can be tailored to specific problems on the ward at the time.

Most nurses have difficulty learning about the care of the dying from formal classroom lectures. This emphasizes the importance of the support team's teaching role when visiting patients on the wards. The value of nurses accompanying the team member has been mentioned. Otherwise, time should be spent talking to the nursing staff after patients have been seen. Where possible, the support team member should supervise rather than carry out practical procedures such as setting up syringe pumps to administer subcutaneous opioids. This should ensure that the support team is not seen as the sole repository of experience and knowledge about terminal care.

5.3. DOCTORS

Stress from patients and families

Doctors have difficulty dealing with patients who have advanced cancer. Disease relapse and death are seen as failures in an age when technological innovations can prolong life. The need for some sense of accomplishment, coupled with a desire to protect patients from distress, will often motivate doctors to take credit for 'curing' cancer with an operation or some other treatment. Buckman (1984) noted that ward rounds and clinics proceed more smoothly and quickly if the probability of future relapse is glossed over with phrases such as 'we have got it in time'. He then describes the patient's anger and anxiety when the disease relapses.

Primary teams often refer patients to the support team because of these relationship difficulties. It is helpful to involve the primary team in dealing with the symptoms and other problems. Encouraging participation by offering a number of ideas and strategies for discussion can reduce the sense of failure; it may even generate a sense of satisfaction when the patient improves. Rarely, some doctors prefer to opt out altogether, and the support team virtually takes over.

Cancer patients with previous psychological difficulties may be branded as 'problem' patients. Pain and other symptoms tend to be re-labelled as anxiety. McCue (1982) reviewed the emotions generated in the doctors: anger, avoidance, fear, and despair. These patients will frequently be referred to the support team, even by primary firms who will not use the team for difficult symptom control problems. It is surprising how often the primary firm becomes more responsive to the needs of the patient as the support team diverts 'problem' behaviours away from the doctors. More rational decision-making is the result.

Doctors' communication with terminally ill patients and families frequently breaks down. On occasions, genuine pressures on time will cause doctors to curtail interviews and avoid picking up patient cues which herald questions. The extra time available to the support team is helpful. By reviewing patients before clinics or ward rounds, medical problems can be selected out and presented to the doctors. More time-consuming psychosocial problems can subsequently be dealt with by the team member. It may also be helpful for the support team doctor to prescribe medications for urgent symptom control when the doctors on the primary team are in operating theatre or casualty.

Occasionally, pressure of time is used to cover up uncertainty and lack of experience. Uncertainty may relate to not knowing how to initiate an interview, or how to deal with the ensuing emotional reactions. The fears that sharing bad news will cause the patient to 'give up' or be distressed are frequently fuelled by the families. It is often easier to give the family more information about the illness and prognosis. The support team can break down these barriers by passing on queries when the patient wants to talk about prognosis, or some other difficult aspect of their illness. The temptation is always to respond immediately and directly, and this may be appropriate on occasions. However, taking up the patient's questions with the primary firm will give them the confidence to respond and will enhance the rapport between the support team and the patient.

Making decisions about treatment of patients with advanced cancer or other life-threatening illnesses can be a major source of stress (Degner and Beaton 1987). The emphasis on teaching the basic sciences at medical college leaves doctors ill prepared for the greater degree of uncertainty in clinical practice. Decisions frequently have to be made on the basis of conflicting or incomplete information. A decision for treatment may involve exposing the patients to considerable risk for marginal benefits. Financial and legal constraints play an increasing role in the decision-making process.

Patients and families are becoming increasingly less satisfied with a passive role in the decision-making process. Patients may want to push for treatments that are unacceptable to the primary team. If the primary team accedes, stress about the risk of toxicity or death results; if they decline, there is the risk of patient anger.

The importance of the support team being familiar with the theoretical basis for decision-making was emphasized in the previous chapter. The primary team will often be comforted when their decision to proceed ahead with what the patient wants is confirmed. Some doctors just appreciate off-loading the uncertainty on to someone prepared to lend a sympathetic ear.

Teaching

Support teams working in teaching hospitals are often asked to teach medical students. Formal lectures may be difficult to fit into already crowded curriculums; the number of lectures will depend on the commitment of the co-ordinators of the teaching programme. Teaching is usually carried out by the doctor on the team but it is valuable to use the clinical nurse specialist, social worker, and chaplain to promote the multi-disciplinary approach.

Lectures should emphasize the needs of patients and that these needs can be met, without necessarily providing a long list of symptoms and their management. It is important to provide guidelines about resources such as books, and about the support team. Most students are keen to learn something about the treatment of pain, nausea, and vomiting because these problems also occur in patients who are not terminally ill. Bart's Medical College now has a minimum standard of learning which requires students to demonstrate an understanding of the basics of pain relief, particularly the use of morphine. This has improved attendance at lectures.

Some teams organize small group sessions about improving communication. The use of role play and video recordings of mock interviews help to prepare students for the difficult job of breaking bad news. Exposing students to dramatic emotional reactions in a controlled situation is particularly useful (Hoy *et al.* 1984). These methods can be extended to include junior doctors in training and members of other disciplines (Nash 1984).

The support team worker should discuss any patients with the medical students who are assigned to them. Students may appreciate accompanying the key worker when their patient is seen. This will supplement their theoretical experience. Some medical students want to spend time with the support team: short visits for one or two days, or longer periods as part of an elective.

After graduation, junior doctors must confront the reality of managing patients. Many become more receptive to learning practical details about pain control. This is a very formative stage in the doctor's career. A good deal of learning is based on experience gained on the job; junior doctors quickly become aware that the preferences of their seniors take precedence over any formal teaching in the past. When faced with a patient complaining of pain or some other distressing symptom, the doctor may occasionally ask the nurses for advice. Usually, they will refer to the senior registrar or consultant, and their advice will become the basis for future decisions. This perpetuates myths about the use of morphine.

The importance of experiential learning should be exploited by the support team. Whenever the team recommends a change in a patient's management, they should discuss the reasoning with the housestaff, preferably in person rather than by phone. The policy of the support team doctor not writing prescriptions means junior hospital staff are more likely to become proficient with the use of drugs such as morphine. The majority of the junior doctors will be grateful for this support and they will often phone or talk about other patients who do not necessarily need to be seen by the team.

Some teams provide formal lectures to health centres and general practice trainees. Invitations to speak at symposia or courses may also be forthcoming. It is much less common to be involved in formal teaching at consultant level. However, the opportunity to give presentations at medical and surgical grand rounds can provide a platform for reviewing issues such as the management of pain, and malignant bowel obstruction.

5.4. OTHER PROFESSIONALS

The needs of other health professionals should be considered by the support team. It is important to get to know the ward clerks. They frequently become the recipients of staff stress, as well as experiencing their own sense of loss when patients die. Medical typists, out-patient clerks, radiographers, and laboratory and other staff will appreciate a personal approach when you want to expedite an appointment or results. Always remember to thank them for their help. This contrasts sharply with the usual impersonal telephone calls they receive.

Paramedical staff such as pharmacists, occupational and physiotherapists, and dieticians find their skills are often undervalued and underutilized. Yet terminally ill patients frequently benefit from their expertise. However, these staff may be unaware of the special needs of terminally ill patients and the sense of urgency required when patients are deteriorating quickly. The key workers should make a special point of discussing relevant problems with these staff when they are involved with patients. The opportunity to share information will enhance their enthusiasm and satisfaction, as well as patient care.

Pharmacists are at a particular disadvantage because they have little contact with patients. There are always reasons for team members to visit the pharmacy; deteriorating patients exhibit greater sensitivity to medication and it is helpful to discuss alternative drugs or routes of administration. If you provide feedback about patient response, particularly improvement, the pharmacists will become motivated to assist with unusual or urgent requests for future patients.

There are three other services with which a support team may be involved: the pain clinic, radiotherapy, and medical oncology. The next chapter reviews the contribution of these services to the care of patients with advanced cancer.

6

The pain clinic and palliative oncology

6.1. INTRODUCTION

The pain clinic and oncology services have an important role in treating pain and other symptoms of advanced cancer. Obviously, the role of the support team will overlap with these services. There is the potential for misunderstanding and conflict; although such services may feel threatened by a support team, the problem is not one-sided. If support team members have not worked in a pain clinic or oncology ward, they may be ignorant of the benefits these services can offer. These team members may want to 'protect' patients from 'aggressive' treatments, and only use symptomatic medical treatment.

To help overcome these problems, we have reviewed the roles of the pain clinic and oncology services. We have provided insights about how oncologists decide to treat patients, and guidelines for working with oncology services.

6.2. PAIN CLINIC

Role of the pain clinic in the management of malignant pain

The commonest symptom experienced by patients with advanced cancer is pain. It has been estimated that in 90 per cent of these patients, satisfactory pain control can be obtained by pharmacological means alone. In the remaining 10 per cent, however, referral to a pain clinic for specialized techniques for the relief of pain may be required. This latter figure may be an underestimate. At one hospice which was visited on a weekly basis by a pain clinic anaesthetist, 22.7 per cent of the patients admitted underwent a specialized pain-relieving procedure (Saunders 1986). This emphasizes the need for support teams to establish and maintain close links with a pain clinic or an anaesthetist with a special interest in nerve blocks.

General principles

The assessment of the patient's pain, its cause, and the patient's reaction to it are an important function of the pain clinic and an essential preliminary to successful management (Baxter 1984). In particular, the location, characteristics, and temporal factors of the pain must be noted and a full neurological examination of the relevant area documented. If possible, the examiner should try to distinguish between local peripheral pain and referred pain, and between somatic and autonomic pain, as well as documenting which nerve or nerves are involved. The anticipated procedure and any potential complications must be discussed with the patient, the referring medical and nursing staff, and, if possible, the patients' relatives.

Before embarking on a permanent nerve block it is advisable, if practicable, to perform a diagnostic local anaesthetic block to enable both the patient to experience the likely benefits of a permanent block and the doctor to confirm the accuracy of the diagnosis (Charlton 1986).

Classification of available techniques

Specialized techniques for the relief of pain can be classified into the following two groups.

1. Techniques which interrupt the pain pathway. The pain pathway can be interrupted peripherally (i.e. with a 'nerve block') or centrally. These techniques are the most commonly used for the control of malignant pain.
2. Techniques which relieve pain by stimulation of the peripheral and central nervous systems. The place of stimulation techniques (e.g. acupuncture, transcutaneous nerve stimulation, and implanted stimulators) in malignant pain has not yet been fully assessed.

Nerve blocks

The commonest indications for a nerve block are:

(1) localized pain breaking through otherwise adequate analgesia;
(2) pain controlled at rest, but not on movement;
(3) attempt to reduce analgesic dose;
(4) failure of analgesics to control pain.

Methods used for nerve blocks

The pain pathway can be interrupted by the following methods.

1. Neurolytic agents: phenol and alcohol are the most frequently used. The great advantage of this method is that a minimum of equipment is required, and in many instances they are an adequate substitute for major neurosurgical procedures. However, these chemicals destroy nerve fibres indiscriminately, and therefore accurate and discrete placement of the agent is essential.
2. Cryolesion: the cryoprobe produces an ice ball at the probe tip. The nerve is placed within the ice ball and repeated freeze–thaw cycles cause Wallerian degeneration and axonal disruption. This technique causes fewer complications than neurolytic agents, but the duration of analgesia is very variable, from a few days to several months.
3. Radiofrequency thermocoagulation: with the aid of a stimulating current, an electrode is accurately localized to the appropriate nerve. By means of a high-frequency electrical current, thermocoagulation of the nerve elements occurs. This technique has the advantage of a low incidence of complications and is superceding the use of neurolytic agents on peripheral nerves. The procedure does, however, require considerable expertise and the equipment is expensive.
4. Spinal opiates: the discovery of opiate receptors in the spinal cord has led to the administration of opiates by the intrathecal or epidural routes. Totally implantable systems are now available for long-term use. A major advantage of spinal opiates over local anaesthetic agents is that they do not cause motor blockade or hypotension (Baxter 1984). There is, however, a small risk of respiratory depression with this method.

Spinal neurolysis

The spinal nerve can be blocked at three different sites.

1. Subarachnoid (intrathecal) neurolysis: the injection of a neurolytic agent onto a spinal nerve requires few special facilities and equipment and this technique can be made widely available. Useful analgesia is experienced in 70 per cent of cases. The most serious complications are motor paresis and interference with bladder and rectal sphincters (Bonica 1953). These complications can be minimized by careful selection of patients and immaculate technique.
2. Epidural neurolysis: this theoretically offers the advantage of fewer complications than the subarachnoid route, but the analgesia is not usually as effective.
3. Subdural neurolysis: this procedure can be particularly useful for treating pain in the distribution of the cervical nerve roots, but the technique is complex and requires x-ray facilities (Lipton 1979).

Peripheral nerve blocks

In general, peripheral nerve blockade produces relatively short-lived pain relief, as compared with spinal neurolysis. Radiofrequency thermocoagulation of peripheral nerves is, however, more effective than peripheral chemical neurolysis. Peripheral nerve blocks are most commonly used for pain in the head and neck, and thorax.

Autonomic blocks

There are many instances in which some or all of the cancer pain is due to involvement of autonomic nerves and for which block of this system is indicated. Any pain that is characterized as burning and in which hyperpathia is present warrants a trial of sympathetic blockade (Charlton 1986).

The most commonly used autonomic blocks are as follows.

1. Coeliac plexus block: produces prolonged relief of abdominal and back pain in over 90 per cent of patients with pancreatic and gastric cancer (Baxter 1984).
2. Lumbar sympathetic block: pain from carcinoma of rectum, bladder, or uterus may be helped by lumbar sympathetic blockade. Some patients with malignancy in the pelvic region

develop burning pain in the lower limbs and this can be relieved by lumbar sympathetic neurolysis.
3. Stellate ganglion block: in the upper thorax, tumour spread sometimes involves sympathetic fibres causing oedema, cyanosis, and a burning pain in the arm. In such cases, stellate ganglion block will be valuable. Other conditions where consideration should be given to a stellate ganglion block are in scar pain following a radical mastectomy and in Pancoast's tumour.
4. Intravenous regional guanethidine block: guanethidine produces a sympathetic block by displacing noradrenalin from stores in sympathetic nerve endings. This technique can be employed for pain in the upper and lower limbs.

Upper limb

Peripheral nerve blockade of the upper limbs is complicated by a high risk to motor function. However, brachial plexus neurolysis should be considered in patients with carcinomatous involvement of the brachial plexus, Pancoast's tumour, and pathological fractures of the upper limb (Charlton 1986).

Subdural neurolytic blocks can be very useful for shoulder pain, with a success rate of 70 per cent (Baxter 1984). However, the analgesia is often short-lived and the technique is complex.

Some arm pain may have both a sympathetic and somatic component, and sympathetic blockade either by stellate ganglion block or by intravenous regional guanethidine block may be of benefit (see above).

Lower limb

Subarachnoid (intrathecal) block is the most useful nerve block for pain in the lumbosacral root distribution, with an overall success rate of over 70 per cent. The complications of motor paresis and bladder and rectal dysfunction can be kept to less than 2 per cent by careful selection of patients and scrupulous attention to technique.

The regional hip block, performed by local anaesthetic blockade of the obturator nerve and nerve to quadratus femoris (Lipton

1979) can last several months in patients with acetabular or femoral lesions. This procedure carries virtually no risks.

Lumbar sympathetic blockade should always be considered for patients with burning pain and hyperpathia in the lower limb (see above).

Head and neck

The most commonly performed peripheral nerve block in the head and neck is of the trigeminal ganglion, and this block can be useful for the control of pain from advanced orofacial malignancy (Charlton 1986). Block of the whole trigeminal nerve may not be necessary as orofacial malignancies usually involve only the lower two divisions and the maxillary and mandibular nerves are amenable to nerve blockade.

Glossopharyngeal and vagus nerve blocks may be needed when pain is arising from the pharynx, larynx, and related structures, and subdural neurolysis is useful for pain in the distribution of the cervical roots (Charlton 1986).

Thorax

Paravertebral and intercostal nerve blocks are useful for pain due to small metastases, pathological rib fractures, and breast cancer. However, these peripheral nerve blocks are most effective if limited to no more than 2 or 3 roots (Charlton 1986). More widespread pain is better managed by epidural neurolysis or spinal opiates.

Abdomen and pelvis

Coeliac plexus block can relieve pain in over 90 per cent of patients with pancreatic and gastric cancer and patients should undergo this block as soon as conventional analgesics are inadequate. An alternative to coeliac plexus block is the splanchnic nerve block. Complications from both these procedures are few.

Perineal pain is best relieved by subarachnoid (intrathecal) neurolytic block, although this procedure carries some risk to bladder and rectal function. Useful alternatives to subarachnoid neurolysis, such as transsacral nerve block and caudal epidural cryolesion, have been described but, although the risks to sphincter function with these techniques is negligible, pain relief by these methods is usually of short duration.

Pelvic visceral pain due to rectal, bladder, or uterine cancer is often relieved by lumbar sympathetic blockade.

Cordotomy

Cordotomy is the most common central method used for interrupting the pain pathway.

The anterolateral quadrant of the spinal cord can be sectioned surgically (open cordotomy) or, more commonly nowadays, percutaneously with a radiofrequency probe (percutaneous cervical cordotomy). Cordotomy should be considered for:

(1) unilateral cancer pain;

(2) pain below C5 dermatome;

(3) pain of well-defined dermatomal distribution;

(4) short life expectancy.

The main complication of a cordotomy is motor weakness in the ipsilateral lower limb; 20 per cent of patients experience difficulty in walking after the procedure. However, over 80 per cent of patients obtain total relief of unilateral pain following a percutaneous cervical cordotomy (Lipton 1979).

Pituitary ablation

A trans-sphenoidal approach to the gland is made and the pituitary is destroyed either with a small volume of alcohol or by the application of a cryoprobe.

This technique is particularly useful for diffuse pain due to widespread bony metastases from hormone-dependent tumours (breast, prostate, kidney, thyroid), but has also been used, empirically, with success for non-hormone-dependent malignant pain when other methods have failed or are unavailable. The mechanism of action is unknown.

Of patients with hormone-dependent tumours, 70 per cent are relieved of their pain, but the procedure does carry a mortality of 5 per cent, and most patients develop diabetes insipidus (Lipton 1979).

Stimulation techniques

Little documented evidence exists on the usefulness of acupuncture in malignant pain, but it would appear not to be generally effective in the majority of patients.

The place of transcutaneous nerve stimulation and implanted stimulators (spinal cord stimulation, deep brain stimulation) in malignant pain is unknown, but transcutaneous nerve stimulation and spinal cord stimulation do not appear to be very promising. Deep brain stimulation can provide excellent pain relief in certain patients, but the procedure requires special facilities and is expensive.

Hypnosis

Although hypnosis has been used in some units with varying degrees of success (Baxter 1984), it would seem unlikely that this technique will replace more conventional methods of treating malignant pain.

Non-malignant pain

Patients with cancer are as likely as the remainder of the population to suffer from non-malignant pain. Referral of a patient, suffering from malignant disease, to a pain clinic, for advice and management of non-malignant pain may be appropriate.

Domiciliary procedures

The introduction of domiciliary care into the management of cancer has produced welcome improvements in patient care.

Although some procedures for the relief of pain could theoretically be performed in the patient's home, it would seem prudent to restrict domiciliary procedures to only the most exceptional of circumstances.

6.3. PALLIATIVE ONCOLOGY

In general, radiotherapy and chemotherapy are administered in designated regional centres. Occasionally, hormone therapy and less toxic chemotherapy may be prescribed by surgeons and physicians on general wards. A significant percentage of cancer patients will be considered for these treatments. It therefore behoves support teams to be aware of the principles and practice of anti-cancer treatment, particularly if working in a hospital that serves as a regional centre. Even in hospitals without such services, you will become involved with patients who would be appropriate for referral, or be asked to provide local follow-up for patients on active treatment elsewhere.

We have not tried to provide an in-depth description of these treatment modalities, and there is considerable variation in the practice thereof from unit to unit. Nevertheless, some basic concepts will be outlined which will facilitate shared care, particularly if you have not had previous oncology experience.

The first important concept is the meaning of the word 'palliative'. In the hospice setting, 'palliative' refers to the control of symptoms by the use of medications and non-pharmacological interventions. If the underlying cause is the cancer, it remains unchecked. This meaning is closest to the latin root 'pallium'—a cloak (OED). Oncologists use the term 'palliative' to refer to treatment which slows down or shrinks the cancer at one or more sites for as long as possible, with no expectation of cure. Symptoms are relieved for as long as the cancer remains in check. The latter meaning will pertain throughout this chapter.

6.4. RADIOTHERAPY

Radiotherapy is the administration of various types of radiation, either from machines outside the body, or applied directly into or close by a cancer. It has a high chance of curing some localized cancers such as Hodgkin's lymphoma, laryngeal carcinoma, and cervical carcinoma.

Radiotherapy can also facilitate the symptomatic management of haemoptysis and cough due to endobronchial tumour, metastatic bone pain, and selected cases of spinal cord compression.

Patients with headaches and neurological deficits from intracerebral tumours or metastases may be considered for treatment. Control of rectal discharge or recurrent vaginal bleeding may be achieved for advanced inoperable pelvic cancers. Some doctors will treat pain associated with advanced head and neck tumours, dysphagia from advanced oesophageal cancer, and extensive soft tissue infiltration or ulceration, particularly from advanced, locally recurrent breast cancer.

A histological diagnosis of the cancer is almost always a prerequisite for treatment. If a diagnosis has not been made, an operation or other biopsy technique will be used. This principle may be waived if the patient has advanced disease, or abnormal non-invasive tests are consistent with the expected clinical pattern of metastatic disease and the risk of the biopsy is too great.

Efforts will frequently be made to identify how widespread the cancer is—a process known as 'staging'. This may seem to expose the patient to a number of unnecessary tests but if the cancer is found to have spread to other important organs, such as liver or lung, lower doses of treatment may be used to reduce toxicity, or systemic hormone or chemotherapy may be considered. If the intention is merely to palliate the cancer at one site, staging is unnecessary. Even then, investigations may reveal other potential trouble spots such as impending fractures. These can then be treated before the patient develops problems.

Patients who have very widespread disease and who are very ill will often not be considered for treatment. They find it difficult to be transported even short distances and lie still enough for treatment. Furthermore, they may not live long enough for the radiation to have a clinical effect.

When a decision is made to give radiotherapy, the treatment is carefully planned. Considerable care is needed to minimize radiation damage to normal tissues, especially in areas previously treated. Treatment is usually given by a number of doses (fractions) over a period of time, ranging from days to weeks. This reduces side-effects and, theoretically, increases the effect on the cancer. Some radiotherapists prescribe short courses, even single doses, when they are palliating symptoms in patient with advanced cancer who are relatively weak and cannot tolerate prolonged courses of treatment. Painful bone metastases are often treated this way. Long-term effects are unlikely to occur if the prognosis is short (Arnott 1987). Not all therapists agree with this approach and it is important to be aware of the attitudes of individual radiotherapists.

In general, the side-effects of radiation treatment are now less frequent and less severe than in the past because of the advances made in the delivery of radiation, and the understanding of radiobiology. Many older patients will need reassurance because they will recall friends or relatives who suffered from treatment given many years ago. Acute effects include nausea and vomiting, particularly when the bowel is irradiated. Diarrhoea may also be a problem from abdominal or pelvic radiotherapy. Inflammation and dryness of the mouth can be especially troublesome during treatment of head and neck tumours, as can the burning retrosternal pain of oesophagitis when the mediastinum is treated. Some degree of alopecia, which is usually reversible, is invariable with cranial irradiation.

The delayed effects of radiotherapy such as bone necrosis, non-healing ulceration, spinal cord transection, and blindness, are now very uncommon. It is important to be aware of late effects which mimic the symptoms of advanced cancer and lead to patients being referred for terminal care. Fistula formation and cachexia from malabsorption after pelvic radiotherapy are examples. The pain of radiation-induced nerve damage may mimic malignant infiltration. The breathlessness and cough of post-radiation pneumonititis may be difficult to distinguish from the symptoms of primary lung cancer or lymphangitis carcinomatosis.

6.5. HORMONE AND CHEMOTHERAPY

Hormone therapy is the use of hormones, such as oestrogens or progestagens, or drugs which block the production or effect of hormones, such as aminoglutethamide and tamoxifen. The growth of some cancers will be affected by hormone manipulation. Chemotherapy refers to the use of drugs which can damage cancer cells directly.

The patients who present for chemotherapy represent a select group. Elderly patients, and patients with concomitant major illness are usually not referred. The philosophy of the primary medical or surgical team which made the diagnosis will be a preselection factor. Many physicians and surgeons recall the toxicity of earlier treatments and will make an arbitrary decision not to refer. Many patients will accept such advice but a percentage will not, particularly the younger patients. A recent study by Slevin *et al.* (1988) revealed that patients referred for chemotherapy will accept a less than 1 per cent chance of cure or relief of symptoms, despite potentially toxic treatment. Patients were prepared to accept a slimmer chance than the professionals who were caring for them.

When a patient first presents for consideration of hormone or chemotherapy, a diagnosis will usually have been made by means of a biopsy and examination of cancer tissue (histology). This prevents patients with non-malignant disorders from receiving potentially life-threatening treatment. To some extent, the histological diagnosis predicts the response of the cancer and this information plays an important part in the plan of the oncology team. A simple guideline of the relative chances of cure and palliation for the more common forms of cancer is given in Table 2. This guideline does not imply that chemotherapy is the treatment of choice for these cancers in every situation.

To make decisions about treatment, oncologists use information about tumour response rates to various types of chemotherapy which is based on a large number of carefully performed studies. They also rely on anecdotal experience about patients who have survived despite the odds. Scenes of mutual celebration often occur when these patients attend clinics. These patients form a major counterpoise to the psychological effects of the deaths of so many others, and their example serves as an incentive for the doctors to encourage similar patients in the future. It is difficult to appreciate the importance of this experience when you first come into contact with an oncology service, particularly when you are primarily involved with the terminally ill patients.

The side-effects of treatment are another factor in the decision-making process. The majority of chemotherapy drugs will affect normal body tissues. Suppression of bone marrow function is common but usually reversible. Anaemia is slow to develop and can be easily treated with transfusions of red cells. A fall in the number of white cells and/or platelets may occur within days of treatment and is potentially far more serious. Patients may die of overwhelming infections, or bleeding into the brain or bowel.

Table 2. Guidelines of relative chances of cure and palliation for the more common forms of cancer

---

Definite chance of cure; good chance of palliation
- leukaemia (some types of childhood leukaemia)
- lymphoma (Hodgkins and some types of non-Hodgkins lymphoma)
- teratoma

Small chance of cure, good chance of palliation
- ovarian carcinoma
- leukaemia (adult)
- lymphoma (most types of non-Hodgkins lymphoma)

Little/no chance of cure; good chance of palliation
- breast carcinoma
- prostatic carcinoma
- small-cell carcinoma lung

Little/no chance of cure; small chance of palliation
- colorectal and stomach carcinoma
- pancreatic carcinoma
- non-small-cell carcinoma lung
- cervical carcinoma

Impairment of immunity may render patients more likely to infections such as shingles.

The commonest side-effects of chemotherapy are nausea and vomiting. These effects may persist for several days after treatment. Hair loss is also quite common and very distressing. Patients are also very distressed by the thought of coming for treatment, the time taken for treatment, and fear of needles (Coates *et al.* 1983). Some patients will experience general effects such as lethargy and anorexia for several months after treatment. Individual agents may produce specific side-effects, such as peripheral neuritis caused by vincristine.

When there is a definite possibility of cure, doctors and patients will accept more side-effects and a greater risk of dying from the treatment. However, one of the major advances in oncology has been the recognition that, in a proportion of cancers, treatment toxicity can be reduced without jeopardizing response rates. This has resulted in shorter courses of treatment with fewer side-effects.

Another factor that is weighed up is the age of the patient. Older patients tolerate chemotherapy less well and are more likely to have other major illnesses. Younger patients may have a greater desire to pursue treatment at any cost, and the plight of these patients, particularly if they have young families, frequently motivates a more aggressive response.

After considering the aforementioned factors, oncologists make one of the three decisions.

1. No treatment. If a decision is made that no treatment is available or should be given to the patient, the patient will usually be discharged.
2. No treatment but observe. This decision is usually made when patients have no symptoms and there is a poor chance of the disease responding to known treatments. Patients may be offered the option of review in out-patient clinic until symptoms develop.
3. Proceed with treatment. This decision may include a distinction between an attempt to cure or to palliate the cancer.

When a decision is made for treatment, further tests are usually done to stage the cancer and measure the size of cancer deposits. The results serve as a baseline to assess if treatment is effective, and may influence the type of treatment to be used. Many investigations, such as blood tests, are carried out routinely, often according to a pre-planned protocol. In a situation where there is a distinct probability of cure, investigations may involve more discomfort and greater risk to patients. Additional investigations may be ordered depending on patient symptoms or unexpected findings from routine tests.

The number and nature of investigations may be reduced if the patient is too unwell or if a life-threatening delay in treatment might result. Where treatment is palliative with a relatively small chance of response, tests will often be kept to a minimum. Oncology units which research and compare various treatment protocols will rarely modify a rigid programme of investigations. This policy may seem callous and inflexible but it has resulted in the considerable improvements in the quality of life afforded to patients receiving palliative treatment.

Most units will have standard treatments depending on the diagnosis and stage of the disease. Some chemotherapeutic agents are used continuously. The majority of cytotoxic drugs are given as a course of injections and/or tablets which are repeated at regular intervals.

Once a patient has been commenced on treatment, they are usually subject to ongoing reviews. If the cancer is physically apparent, the tumour will be measured before each course of treatment is due. After two or three courses of treatment, tests which demonstrate the size of the cancer will be repeated—it often takes several weeks before a response is apparent.

If the cancer is responding, therapy will be continued for a set number of treatments or until the cancer is not detectable. If the cancer progresses through treatment, another aggressive protocol will be used if there is a chance of cure. If this is not feasible or appropriate, therapy will often de-escalate to less intensive palliative treatment. Severe side-effects may lead to a change in, or cessation of treatment, even when the disease is responding.

Response to treatment is not always reviewed. In some instances, such as metastatic bone disease, objective response will lag several weeks behind subjective improvement. It may not be possible to quantitate response in some patients. On other occasions, doctors will continue with treatment without attempting to assess disease because they do not want to know if the cancer is progressing; they want to maintain the patient's hope in the treatment.

The majority of oncology units will provide supportive medical care throughout treatment. Patients and families are educated about symptoms which herald low platelet or white cell count. Regular blood tests will be performed between courses. However, the oncology team may not be as aware of distressing symptoms, such as nausea and vomiting, which are not life-threatening. Patients will frequently underestimate the severity of symptoms which occur between courses.

Increasing attention is being paid to the emotional support of patients and their families. Nursing staff and other patients play an important role in this regard. Some units employ oncology nurse specialists or social workers. Support groups—informal meetings of patients, families, and staff—are also used. Formal psychological support services may also be available.

Patients receiving chemotherapy will frequently be given more information about their disease and the treatment than is usually the case. Doctors assume that patients require to know their diagnosis and treatment prospects in order to cooperate fully with potentially toxic therapy. But it is interesting to note how many patients who are not told their diagnosis will submit to such treatment (Gould and Toghill 1981). Doctors present information at interviews and ward rounds. Oncology nurse specialists will supplement this knowledge. Information booklets may also be made available to patients and families.

6.6. WORKING WITH ONCOLOGY SERVICES

Oncologists and radiotherapists will have many patients who are likely to die despite treatment, and the support team is often asked to be involved with these patients. The following guidelines are offered to facilitate relations between oncology services and the support team.

1. Determine if you can work alongside 'active' treatment. Some teams have tried to help patients to accept impending death rather than pressing on with further 'futile' courses of treatment. Many patients have been angered by this approach and transferred their anger to the primary team for asking the support team to be involved. This has made the oncology team reluctant to refer other patients.

Some patients are referred because of deterioration during treatment, before a response is apparent. The doctors will often want to continue therapy, and this decision to refer will cause them considerable stress (Degner and Beaton 1987). This stress will be compounded if the team member seeks to prevent further treatment being given. If the needs of the patient have not been properly ascertained, it may be more prudent to avoid raising such issues. If these situations become too difficult it may be easier to limit the team's role to those patients who are definitely not having further treatment.

2. Recognize skills that oncologists have in terminal care. It is important to avoid any sense of rivalry that may develop when oncology services incorporate basic concepts of symptom relief and supportive care into their clinical practice.

3. Maintain an open approach with patients who are seeking further active treatment or who are being treated. It is important to listen to what the patient wants, and facilitate this as far as is possible. This involves supporting the patient's desire for an oncology opinion if the primary team have dismissed this option. When listening to patients who are receiving treatment, remember that 'I do not like chemotherapy' does not equate with 'I do not want chemotherapy'. Patients often like to share their frustrations about having no options except therapy, but be careful about how you interpret these frustrations.

4. Report information about the patient's quality of life. Maintaining regular contact with out-patients makes one more aware of the incidence and severity of side-effects. This information should be reported either directly to the oncologist or in the medical notes. This will often facilitate better control of symptoms during treatment. The patient may need encouragement to seek readmission if there are life-threatening side-effects.

5. Identify who makes decisions about patient treatment. Decisions relating to the initiation or discontinuation of treatment usually rest with the consultant or senior registrar. One will need to approach them if the patient genuinely wishes to stop therapy. However, decisions about minor modifications in analgesics and other medications for symptom control may be made by the junior doctors.

6. Identify where decisions are made. Decisions about in-patients will often be made or ratified during formal ward rounds. These rounds can be daunting unless you accompany them routinely. Informal ward rounds are a less intimidating forum, particularly for issues pertaining to symptom control.

Combined clinics and meetings bring together several disciplines to discuss problem cases. Where possible, these meetings should be attended; they provide an ideal opportunity to raise points about terminal care, and other teams will become familiar with your work.

7. Be aware of the doctors' individual preferences about treatments. Time should be spent with the nearest oncology unit, during orientation to the support team, becoming aware of the referral pattern of the physicians and surgeons who use these services.

When working closely with a unit, do not presume on one's own familiarity with the different approaches that doctors have to treatment. A relationship with a patient may be jeopardized if an alternative treatment programme is decided upon.

8. Facilitate the transfer of distressing information. Hinton (1979) observed on one radiotherapy unit that there was a relative lack of frank discussion about dying. This did not appear to relate to pressures on time. He postulated that it was due to the emotional investment required when treating patients. This commitment made it difficult for carers to adjust to treatment failure. The support team can be in a better position to pick up on issues relating to disease progression.

It is often thought that patients receiving anti-cancer therapy cannot come to terms with their diagnosis because they are constantly focusing on the hopeful outcome of treatment. A small minority of patients deny their illness, but most will recognize when their cancer is progressing through treatment. They are very aware of lumps increasing in size, deterioration in their general health and specific symptoms, and subtle changes in the way the primary team relates to them.

Although patients are more likely to know their diagnosis and what treatment entails, information about the likelihood and consequences of treatment failure is less forthcoming. This applies particularly when curative treatment is considered, but often applies to palliative therapy. Many patients will want confirmation about the progress of their illness, and it can be helpful to get the primary team to review test results with them. It is gratifying how readily doctors will include this information when they realize that the patient wants it (Reynolds *et al.* 1981). Confirmation that the cancer is progressing will frequently precipitate opportunities to tackle fears and unresolved burdens. The patient will not necessarily want to abandon treatment.

Epilogue

These guidelines should give some insight into the advantages that result from an atmosphere of mutual co-operation. Managing the distress and the symptoms of advancing cancer, and being prepared to work alongside the primary team, will greatly reassure patients. Families and the professional carers will also appreciate the extra dimension of supportive care that can be offered. This will allow you to become part of a truly integrated service; a service offering a balanced approach to patients and families who are struggling to come to terms with the devastating impact of advanced cancer.

> Death must simply become the discreet but dignified exit
> of a peaceful person from a helpful society,
> without pain or suffering, and ultimately without fear.
> Philippe Airies, 1977
> *The Hour of Our Death*

References

Ahmedzai, S. (1982). Dying in hospital: the residents' viewpoint, *British Medical Journal*, **285**, 712–14.

Arnott, S. (1987). The role of radiotherapy in treatment of metastatic cancer. In *Management of metastases*, Balliere's Clinical Oncology—International Practice and Research, Vol. 1, (ed. M. L. Slevin), pp. 537–50. Balliere Tindall, London.

Bates, T. D. (1985). *St Thomas' Hospital Terminal Care Support Team: eighth annual report.* Available from: The secretary, Support Care Team, St Thomas' Hospital, London SE1 7EH.

Bates, T. D., Hoy, A. M., Clarke, D. G., and Laird, P. P. (1981). The St Thomas' Hospital Terminal Care Support Team—a new concept of hospice care. Lancet, **i**, 1201–3.

Baxter, R. (1984). Specialized techniques for the relief of pain. In *Management of terminal disease*, 2nd edn, (ed. C. M. Saunders), pp. 91–99. Edward Arnold, London.

Beachamp, T. L. and Childree, J. F. (1983). *Principles of biomedical ethics*, 2nd edn. Oxford University Press, New York.

Beckhard, R. (1974). Organizational implications of team building. In *Making health care teams work*, pp. 69–98. (ed. Wise, H., Beckhard, R., Rubin, I., and Kyte, A. L.) Ballinger, Cambridge, Massachusetts.

Billings, J. A. (1985). *Outpatient management of advanced cancer*, J. B. Lippincott Company, Philadelphia, Pennsylvania.

Bonica, J. J. (1953). *The management of pain.* Lea and Febiger, Philadelphia, Pennsylvania.

Bolam v Friern Hospital Management Committee. 1 W.L.R. 582.

Buckman, R. (1984). Breaking bad news: why is it still so difficult? *British Medical Journal*, **288**, 1597–9.

Cartwright, A., Hockey, L., and Anderson, J. L. (1973). *Life before death.* Routledge & Kegan Paul, London.

Charlton, J. E. (1986). Current views on the use of nerve blocking in the relief of chronic pain. In *The therapy of pain*, 2nd edn, pp. 133–64. (ed. M. Swerdlow). MTP Press Limited, Lancaster.

Coates, A., Abraham, S., Kaye, S. B., Sowerbutts, T., Frewin, C., Fox,

R. M., Tattersall, M. H. (1983). On the receiving end—patient perception of the side-effects of cancer chemotherapy. *European Journal of Cancer and Clinical Oncology*, **19**, 203–8.

Degner, L. and Beaton, J. I. (1987). *Life-death decisions in health care*. Hemisphere, Washington, DC.

Dunlop, R. J., Hockley, J. M., and Davies, R. J. (1989). Preferred versus actual place of death—a Hospital Terminal Care Support Team experience. *Palliative Medicine*, **3**, 197–201.

Gould, H. and Toghill, P. J. (1981). How should we talk about acute leukaemia to adult patients and their families? *British Medical Journal*, **282**, 210–12.

Greer, D. S., Mor, V., Morris, J. N., Sherwood, S., Kidder, D., and Birnbaum, H. (1986). An alternative in terminal care: results of the National Hospice Study. *Journal of Chronic Diseases*, **39**, 9–26.

Hector, W. (1974). Nursing. In *The Royal Hospital of St Bartholomew*, p. 460. (ed. V. C. Medvei and J. L. Thornton). Covell, London.

Herxheimer, A., Begent, R., MacLean, D., Phillips, L., Southcott, B., and Walton, I. (1985). Short life of a terminal care support team: experience at Charing Cross Hospital. *British Medical Journal*, **290**, 1877–9.

Hinton, J. (1963). The physical and mental distress of the dying. *Quarterly Journal of Medicine*, **32**, 1–21.

Hinton, J. (1979). Comparison of places and policies for terminal care. *Lancet*, **i**, 29–32.

Hockley, J. (1989). Caring for the dying in acute hospitals. *Nursing Times*, **85**, 47–50.

Hockley, J. M., Dunlop R., and Davies, R. J. (1988). Survey of distressing symptoms in dying patients and their families in hospital and the response to a symptom control team. *British Medical Journal*, **296**, 1715–17.

Hoy, A. M., Saunders, B. M., and Kearney, M. (1984). Breaking bad news. *British Medical Journal*, **288**, 1833.

Kristjanson, L. J. (1986). Indicators of quality of palliative care from a family perspective. *Journal of Palliative Care*, **1**, 8–17.

LaGrand, L. E. (1980). Reducing burnout in the hospice and the death education movement. *Death Education*, **4**, 61–75.

Lipton, S. (1979). *The control of chronic pain*. Edward Arnold, London.

Lunt, B. and Hillier, R. (1981). Terminal care: present services and future priorities. *British Medical Journal*, **283**, 595–8.

McCue, J. D. (1982). The effects of stress on physicians and their medical practice. *New England Journal of Medicine*, **306**, 458–63.

MacAdam, D. B. and Smith, M. (1987). An initial assessment of suffering in terminal illness. *Palliative Medicine*, **1**, 37–47.

Maguire, P. (1985). Barriers to the psychological care of the dying. *British Medical Journal*, **291**, 1711–13.

Moore, N. (1918). *The history of St Bartholomew's Hospital*. Pearson, London.

Morris, W. A. (1981). Care of the terminally ill in a district general hospital. *British Medical Journal*, **282**, 287–8.

Mount, B. M. (1980). Personnel selection, applying the McMurray principles to palliative care. In *The R.V. H. manual on palliative/hospice care*, pp. 21–6. (ed. I. Ajemian and B. M. Mount). Arno Press, New York.

Mount, B. M., Jones, A., and Patterson, A. (1974). Death and dying: attitudes in a teaching hospital. *Urology*, **4**, 741–7.

Nash, T. P. (1984). Breaking bad news. *British Medical Journal*, **288**, 1996.

NWTRHA (North West Thames Regional Health Authority) (1987). *Regional strategy—towards a strategy for dying and bereaved people*. Available from: 40 Eastbourne Terrace, London, W2 3QR.

Parkes, C. M. (1985). Terminal care: home, hospital, or hospice? *Lancet*, **i**, 155–7.

Rainey, L. C., Crane, L. A., Breslow, D. M., and Ganz, P. A. (1980). Cancer patients' attitudes toward hospice services. *Ca: A Cancer Journal for Clinicians*, **34**, 191–201.

Reynolds, P. M., Sanson-Fisher, R. W., Desmond Poole, A., Harker, J., and Byrne, M. J. (1981). Cancer and communication: information-giving in an oncology clinic. *British Medical Journal*, **282**, 1449–51.

Saunders, C. (1986). Current views on pain relief and terminal care. In *The therapy of pain*, 2nd edn, pp. 239–59. (ed. M. Swerdlow). MTP Press Limited, Lancaster.

Saunders, C. (1988). The evolution of the hospices. In *The history of the management of pain: from early principles to present practice*. (ed. R. D. Mann). Parthenon Publishing Group, Carnforth, Lancashire.

Sidaway, vs. Bethlem Board of Governors. (1985). 2 W.L.R. 480 (HL).

Slevin, M. L., Plant, H., Lynch, D., Drinkwater, J., and Gregory, W. M. (1988). Who should measure quality of life, the doctor or the patient? *British Journal of Cancer*, **57**, 109–12.

Vachon, M. L. S. (1978). Motivation and stress experienced by staff working with the terminally ill. *Death Education*, **2**, 113–22.

Vachon, M. L. S. (1986). Battle fatigue in hospice/palliative care. In *1986 International symposium on pain control*, pp. 69–76. International congress and symposium series (ed. D. Doyle). Royal Society of Medicine, London.

Vachon, M. L. S. (1987). *Occupational stress in the care of the critically ill, the dying, and the bereaved*. Hemisphere, Washington.

Whitfield, S. A. (1979). A descriptive study of student nurses ward experi-

ence with dying patients and their attitudes towards them. Unpublished Thesis, Manchester University.
Wilkes, E. (1984). Dying now. *Lancet*, **i,** 950–2.
Wright, A., Cousins, J., and Upward, J. (1988). *Matters of death and life. A study of bereavement support in NHS hospitals in England*. King Edward's Hospital Fund for London, London.

Appendix. Recommended reading

Billings, J. A. (1985). *Outpatient management of advanced cancer*. J. B. Lippincott Company, Philadelphia, Pennsylvania.
Copperman, H. (1983). *Dying at home*. John Wiley and Sons, Chichester.
Saunders, C. (ed.) (1984). *The management of terminal malignant disease*, 2nd edn, pp. 232–241. Edward Arnold, London.
Stedeford, A. (1984). *Facing death*. Heineman, London.
Twycross, R. G. and Lack, S. A. (1984). *Therapeutics in terminal cancer*. Pitman, London.
Wilkes, E. (ed.) (1982). *The dying patient: the medical management of incurable and terminal disease*. MTP Press Limited, Lancaster.
Vachon, M. L. S. (1987). *Occupational stress in the care of the critically ill, the care of the dying, and the bereaved*. Hemisphere, Washington.

Index

abdomen, peripheral nerve blocks 80
Accident and Emergency departments 57
accommodation, in palliative care units 13
accountability, pathways of 30
acupuncture 76, 82
acute wards 23
administrative directives 23
advice, communication to primary medical team 53–6
advocacy for patient 53
age of patients for chemotherapy 88
aggression 48, 65
alcohol (neurolytic agent) 77
anaesthetics (regional) 75–81
anger 48–9
  of relatives 9, 10, 66
answer-phones 27, 38
anxiety 7–8
  of nurses 10
  of patients 65
applicants for jobs on support teams 40–1
audit 44
authoritarianism 56
autonomic nerve blocks 78

bad news, *see* prognosis
Barrett, Howard, Dr 3
Bates, Thelma, Dr 15
bequests 26
bereavement iii, 9
  follow-up work 36, 48
biopsy 84, 86
blocks, nerve 76–81
Bolam principle 55
brain stimulation, deep 82
breathlessness and anxiety 8
Brompton Hospital
  Continuing Care Unit 22
  Support Care Team 20–2
Burford, Wendy (nurse specialist in terminal care) 21–2

cancers
  children 18–20
  response to chemotherapy 87 (table)
candidates, for jobs on support teams 40–1

Care of the dying (City and Hackney Health Authority) 37
careers
  doctors on support teams 34
  nurse specialists in terminal care 31
carers, *see* professional carers
casualty departments 57
catering facilities in palliative care units 13
chaplains 36–7
Charing Cross Hospital Terminal Care Support Team 26, 33, 52
charity funding 31
chemotherapy 85–91
  response of cancers 87 (table)
  support of patients on 89–90
  withdrawal 5
chest, peripheral nerve blocks 80
children, cancer 18–20
choices of patients 50–2, 72
'chronic niceness' 60
City of Hackney Health Authority, *Care of the dying* 37
clinical investigations 88
clinical notes 47, 54
clinical nurse specialists, *see* nurse specialists in terminal care
clinical science, 'tough' iii
coeliac plexus block 78, 80
collapse of support teams 42
  Charing Cross Hospital 26, 52
communication
  of advice to primary medical team 53–6
  failure 66, 71
communication of prognosis
  to patients 8, 48–9, 71, 90, 92
  to relatives 9
community care iv, 2
  qualifications 31
  St Thomas's Hospital Terminal Care Support Team 15–16
  see also domiciliary pain relief procedures
community services, conflict with 57–8
compassion, definition iii
complaints 24, 44
compliance, *see* non-compliance
conferences on terminal care 61

Index

conflict
  with community services 57–8
  of nurses with doctors 10, 67
  of oncology services with support teams 90–1
  with primary medical team 53–6
  within support teams 60
consent 50
consulting hospital teams, *see* support teams
Continuing Care Unit at Brompton Hospital 22
continuity of care 45–6
cordotomy 81
counselling skills, workshops 62
cryolesions 77
cures in Middle Ages 2–3
cytotoxic drugs, *see* chemotherapy

Davies, Robert, Dr (St Bartholomew's Hospital) 17
day care, volunteers 39
day care centres, hospice 16
death
  association of support teams with 53
  impending, awareness of 7
  places of 4, 23
decision-making 58–9
  for chemotherapy 88–91
  medical 33, 71
  by patient 51
decor, in palliative care units 13
deep brain stimulation 82
definitions
  compassion iii
  palliation 83
  team 42–3
denial 6, 49–50
depression
  in patients 7
  in relatives 8
despair 50
diagnosis, histological 84, 86
direct telephone lines 27
directives, administrative 23
district general hospitals, deaths 5
doctors
  attitudes to nurse specialists in terminal care 20–1
  conflict with nurses 10, 67
  failure of communication 66, 71
  liaison with 25
  needs 11
  stress 70–3
  support 71–2
  on support teams 31–4
  see also junior doctors
domiciliary pain relief procedures 82
donations 26
dress, nurse specialists in terminal care 30
drivers 38–9
'dumping ground' philosophy 14
dynamics of support teams 42–62

education
  members of support teams 61–3
  pain control 23–4
  see also teaching
Elizabeth Clarke Charitable Trust 21
emotional support, *see* support
emotions 48–9
  see also under specific emotions
epidural neurolysis 78
ethics 50–3
evaluation of performance 44
experience
  for doctors on support teams 32
  for nurse specialists in terminal care 29–3
expert opinions 52–3

facilitators, 'objective' 59
failure, sense of 49
families 34–6
  of children with cancer 18–19
  meeting with 47
  see also relatives
fatigue, relatives 8
fear of support teams 53
feedback 44
finance, *see* funding
first visits to patients 45
follow-up visits 46–7
funding
  conferences on terminal care 61
  Hospital for Sick Children Symptom Care Team 18
  palliative care units 14
  support teams 19, 25–6

Garnier, Mme Jeanne 3
goals of support teams 11–12, 42–4
Goldman, Ann, Dr 19
Great Ormond Street, *see* Hospital for Sick Children
grief 66
group counselling in bereavement 48
group sessions
  and chemotherapy 89
  medical students 72
  nurses 69
guanethidine block, regional 79

head and neck, peripheral nerve block 80
healing in Middle Ages 2–3
health visitors on support teams 30
hip block, regional 79–80
histological diagnosis 84–86
history
  hospice movement 3–4

Index

history (*cont.*)
  support teams 14–22
  terminal care at St Bartholomew's Hospital 2–3
Hockley, Jo, and St Bartholomew's Hospital Terminal Care Support Team 16–17
home, death at 4
home care, *see* community care
home care teams iv
hormone therapy 85
hospice day care centres 16
hospice movement, history 3–4
hospices
  for children 19
  hospital based 23–4
  liaison with 58
  nursing experience 29–30
  transfer to 57
Hospital for Sick Children
  liaison with local hospitals 18–19
  Symptom Care Team 18–20
hospital patients, symptom profiles 6–7
hospitals
  death in 4–6, 23
  transfer from, to hospice 57
hypnosis 82

impending death, awareness of 7
impotence, sense of 49
initial visits to patients 45
interviews, for jobs on support teams 40–1
intrathecal neurolysis 78, 79
intravenous regional guanethidine block 79
investigations, clinical 88
Irish Sisters of Charity 3
isolation, social 39

job descriptions 28
junior doctors
  needs 11
  support 73
  turnover 9, 68

Kerrane, Tom (Chief Nursing Officer of National Heart and Chest Hospitals) 20
key-worker, the 45–6
late referral of patients 48, 57
leader role 58
lectures 58
  to nurses 69
legal considerations 55
limits in team objectives 43
literature on palliative care 55
lower limb, peripheral nerve blocks 79
lumbar sympathetic block 78

Macmillan Cancer Relief Fund 25
malignancy, *see* cancers
management groups 59
managers, nurse 30
measurement of performance 44
medical care in palliative care units 14
medical model 58
medical social workers and nurse specialists in terminal care 21
medical students 64
  teaching 16, 72
medical teams, *see* primary medical teams
medico-legal considerations 55
meetings
  with nurses 68
  of support team 52, 58–9
meperidine 56
Middle Ages, cures 2–3
minority groups 37, 48
morphine, education of doctors 73
multidisciplinary meetings 54, 91

names of support teams 26
Napp (pharmaceutical company) 17
National Hospice Study 23
neck, peripheral nerve blocks 80
needs
  patients 7
  professional carers 10, 63–74
  relatives 8–9, 24
  neoplasms, *see* cancers
nerve blocks 76–81
nerve stimulation 76, 82
neurolysis, spinal 77–8
neurolytic agents 77
night sitting 39
non-compliance 51–2
non-malignant pain 82
notes, clinical 47, 54
number of support teams in United Kingdom 42
nurse specialists in terminal care 20, 28, 29–31
  liaison with 25
nurses
  attitudes to nurse specialists in terminal care 21
  conflict with doctors 10, 67
  needs 10–11
  reduction in number 67
  stress 64–7
  students 63–4, 69
  teaching 69–70
nursing care
  in palliative care units 13–14
  refusal 65

'objective' facilitators 59
objectives, team 11–12, 42–4
office space 26

Index

oncology services 75, 83–92
  paediatric 18
  relationships with support teams 90–2
opiates, spinal 77
opinions, expert 52–3
opposition to support teams 23
out-patient contact with patients 47
overload, role 46

paediatric oncology services 18
pain, non-malignant 82
pain clinics 75–82
pain control, education 23–4
palliative care teams, *see* support teams
palliative care units 2, 13–14, 22
  see also under names of hospitals
palliative oncology 83–92
paramedical staff 74
Parkes, C. M., on terminal care 24
part-time working for doctors on support teams 31–2
paternalism 55
patients
  advocacy for 53
  anxiety 65
  choices 50–2, 72
  communication of prognosis to 8, 48–9, 71, 90
  depression 7
  needs 7
  out-patient contact 47
  'problem' 70–1
  referral, *see* referral of patients
  telephone contact 47
  visiting 39
  visits, initial 45
pelvis, peripheral nerve blocks 80
performance, measurement 44
peripheral nerve blocks 78, 79–81
pethidine 56
pharmacists 74
phenol 77
phones, *see* telephones
pituitary ablation 81
post-mortem, requests 9
prescriptions 54
primary medical teams
  communication of advice 53–6
  compromises 56
  conflict with 53–6
  influence on choices of patients 52
'problem' patients 70–1
professional carers
  liaison with 25
  needs 10, 63–74
  stress 63–74
prognosis, communication
  on chemotherapy units 92
  to patients 8, 48–9, 71, 90
  to relatives 9

property bags 9
protocols, chemotherapy 88–9
psychodynamics of support teams 42–62
psychological needs, *see* needs

qualifications
  community care 31
  nurse specialists in terminal care 29
questionnaires 44

radiofrequency thermocoagulation 77
radiotherapy 83–5
  withdrawal 5
readmission of terminally ill patients 57
record-keeping 44
recruitment 39–41
referral of patients 42, 44–5
  late 48, 57
  by nurses 68
  to St Bartholomew's Hospital Terminal Care Support Team 17
reforming spirit, need for 22–3
refusal of nursing care 65
regional administrative directives 23
regional guanethidine block 79
regional hip block 79–80
relatives
  communication of prognosis to 9
  inability to cope 6
  needs 8–9, 24
  see also families
religions
  non-Christian, spiritual advisers 37
  see also chaplains
research 24
  and doctors on support teams 33
respite care 14
responsibility, pathways of 30
reviews 59
role overload 46
Royal Charter of St Bartholomew's Hospital 3
rules of St Bartholomew's Hospital (1900) 51
Rupert Foundation 19

St Bartholomew's Hospital
  rules (1900) 51
  terminal care, history 2–3
  Terminal Care Support Team 1, 16–17, 24
St Bartholomew's Medical College 72
St Christopher's Hospice 3–4, 24
  conference on terminal care 61
St Joseph's Hospice 3
St Luke's Home for the Dying Poor 3
St Luke's Hospital, New York 15
St Thomas's Hospital Terminal Care Support Team 15–16, 38
science, tough clinical iii

Index

secretaries on support teams 37–8
selection of members of support teams 28–41
Shorter Oxford English dictionary, definition of compassion iii
side-effects
  chemotherapy 86–7
  radiotherapy 85
sisters, ward, attitudes to nurse specialists in terminal care 21
sitting, night 39
small group sessions, *see* group sessions
social gatherings 61
social isolation 39
social needs of relatives 9
social system 35
social workers
  liaison with 25
  and nurse specialists in terminal care 21
  on support teams 34–6
socializing with nurses 68
spinal neurolysis 77–8
spinal opiates 77
spiritual advisers
  non-Christian religions 37
  see also chaplains
spiritual support 8
staff, palliative care units 13–14
staging 84
stellate ganglion block 79
stimulation, nerve 76, 82
stress
  doctors 70–3
  professional carers 63–74
  support teams 48–50, 59–60
student nurses 63–4, 69
subarachnoid neurolysis 78, 79
subdural neurolysis 78, 79
succour iii
supervisory role of support team 70
support
  doctors 71–2
  junior doctors 73
  nurses 64–70
  patients on chemotherapy 89–90
support care teams, *see* support teams
support teams iv–v
  history 14–22
  number in United Kingdom 42
  see also under names of hospitals
symptom control teams, *see* support teams
symptom profiles of hospital patients 6–7
symptoms in children 18

teaching
  by chaplains 37
  by doctors on support teams 33
  of medical students 16, 72
  by nurse specialists in terminal care 30–1
  for nurses 69–70
  visits 58
  see also education
teaching hospitals, deaths 5
team, definition 42–3
team reviews 59
telephones 26–7, 38
  contact with patients 47
terminal care support teams, *see* support teams
thermocoagulation 77
thorax, peripheral nerve blocks 80
training, *see* education
transfer, from hospital to hospice 57
transport 38–9, 47
trigeminal ganglion block 80
tumours, *see* cancers
turnover
  doctors on support teams 34
  junior doctors 9, 68

uniforms 30
United Kingdom, number of support teams 42
upper limb, peripheral nerve blocks 79

Vicary, Thomas (16th-century surgeon) 2–3
visionaries, need for 22–3
visits, teaching 58
visits to patients 39
  follow-up 46–7
  initial 45
volunteers 38–9

ward clerks 73
ward rounds 54, 91
ward sisters, attitudes to nurse specialists in terminal care 21
wards, acute, and terminal care 23
withdrawal
  of chemotherapy 5
  or radiotherapy 5
workload 46
workshops for counselling skills 62